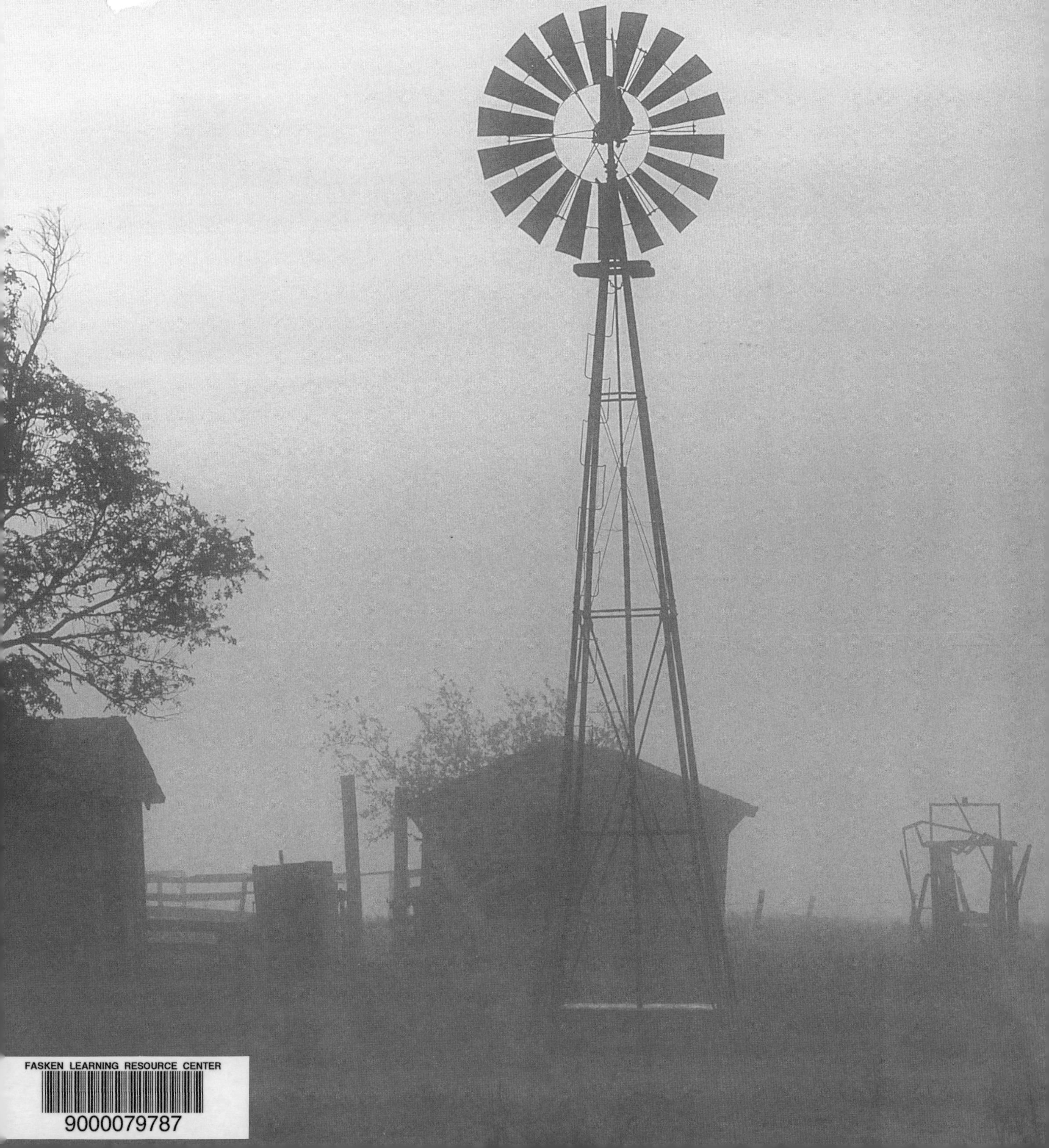

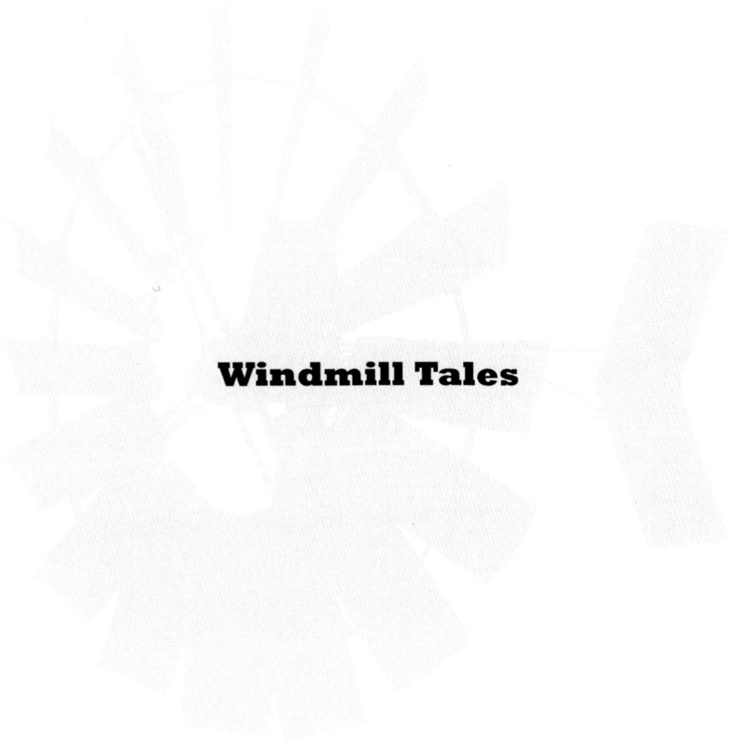

Windmill Tales

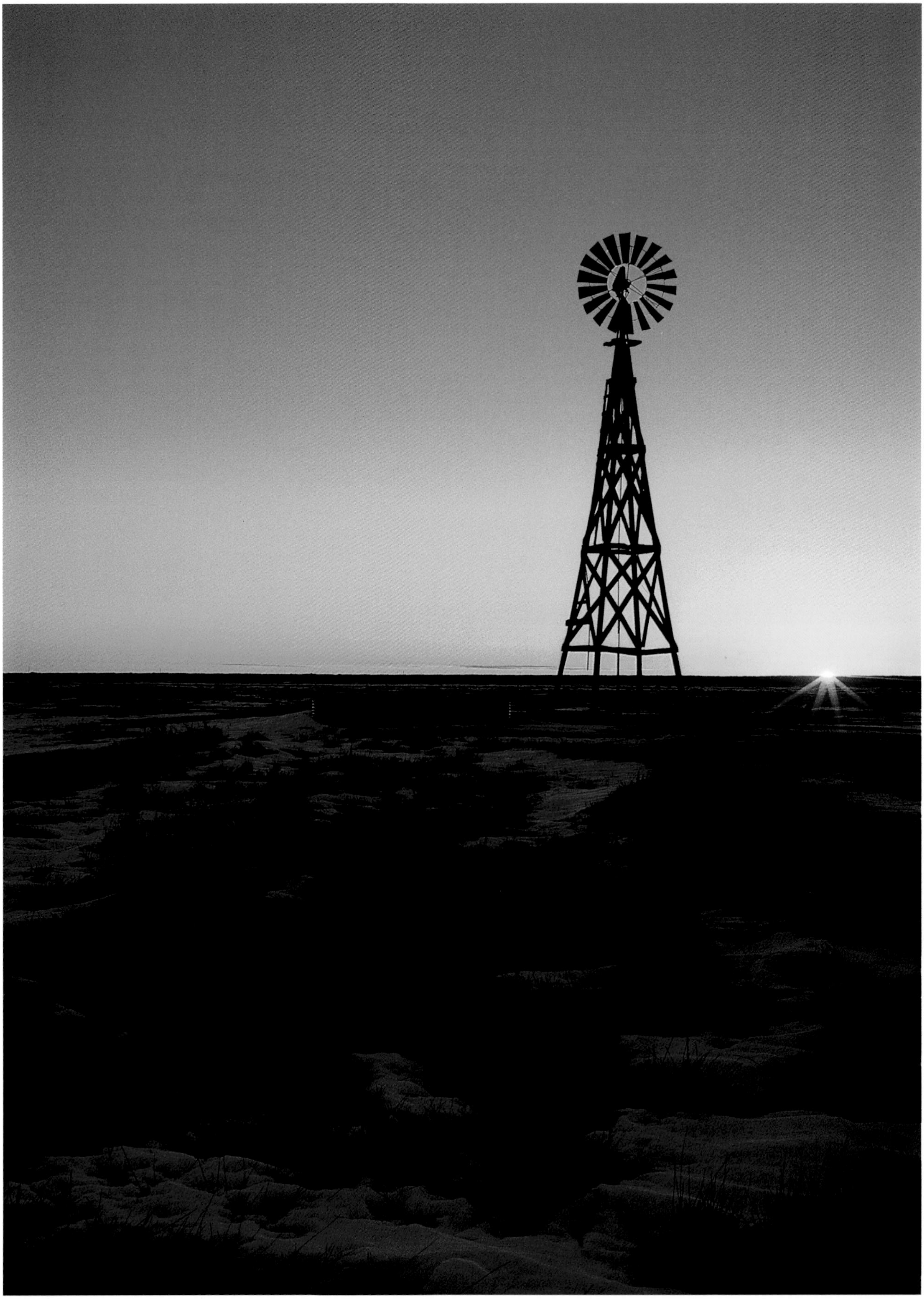

Windmill Tales

Stories from the
American Wind Power Center

PHOTOGRAPHS BY Wyman Meinzer

EDITED BY Coy F. Harris

FOREWORD BY Steve Halladay

Published for the American Wind Power Center
by
Texas Tech University Press
Lubbock

OTHER BOOKS BY THE AMERICAN WIND POWER CENTER

The 702 Model Windmill: Its Assembly, Installation and Use, by T. Lindsay Baker, 1999

Windmills of the World, by William McCook, 2003

This book is typeset in Centaur. The paper used in this book meets the minimum requirements of ANSI/NISO Z39.48-1992 (R1997). ∞

Designed by David Timmons

Printed in Korea

Library of Congress Cataloging-in-Publication Data is available.

04 05 06 07 08 09 10 11 12 / 9 8 7 6 5 4 3 2 1

Texas Tech University Press
Box 41037
Lubbock, Texas 79409-1037 USA
800.832.4042
ttup@ttu.edu
www.ttup.ttu.edu

Publishing of *Windmill Tales* was made possible by a grant from the
PERRY FOUNDATION,
Albuquerque, New Mexico, Susan L. Perry, President.

Proceeds from the sale of this book benefit the
AMERICAN WIND POWER CENTER
in Lubbock, Texas.

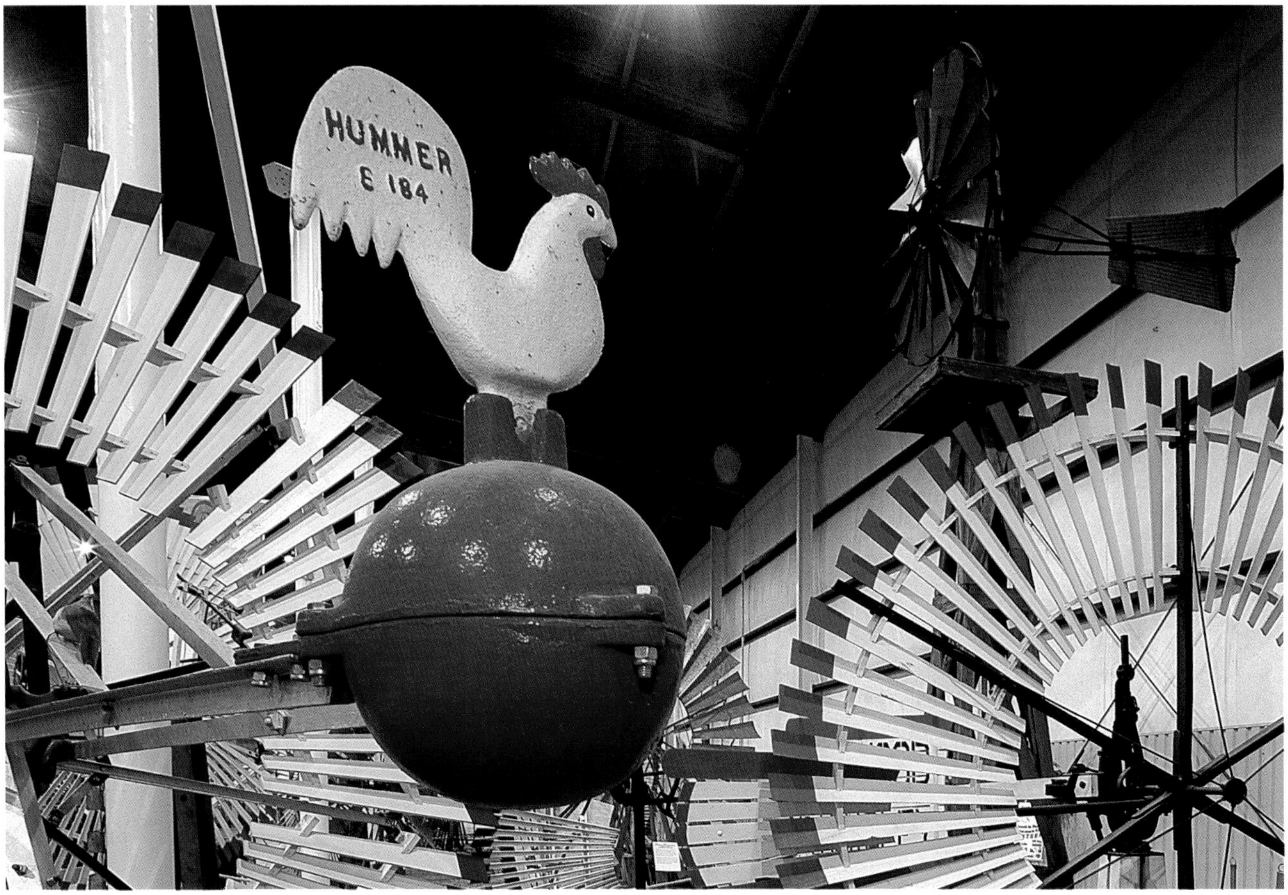
HUMMER
E 184

Elgin Hummer Rooster Weight

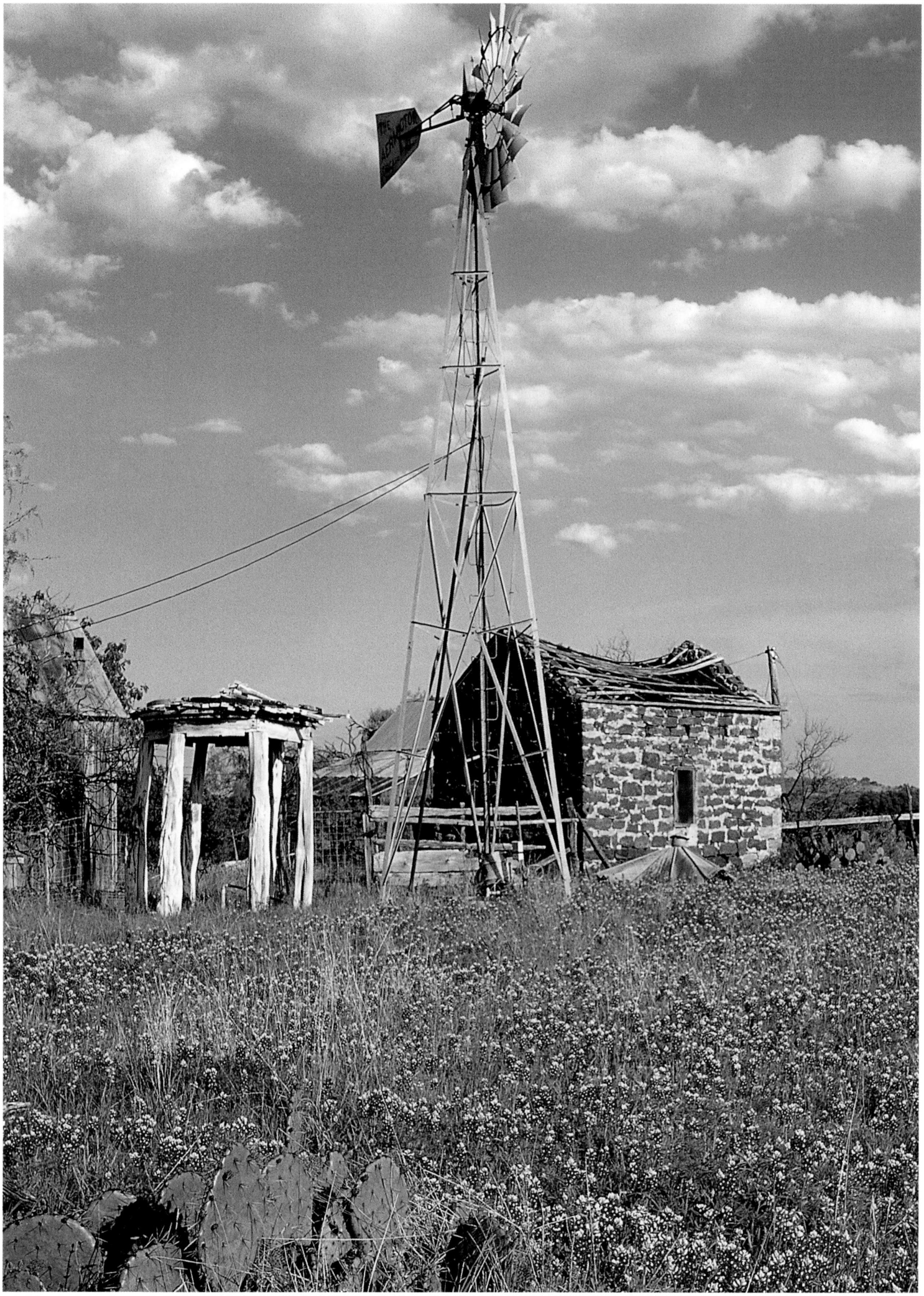

This book is dedicated to
DR. GROVER E. MURRAY
1916–2003

*For his enlightenment of the public to the importance and beauty of
arid and semiarid lands, where the windmill was most needed
and used, and whose advocacy of the windmill museum
was instrumental in its creation.*

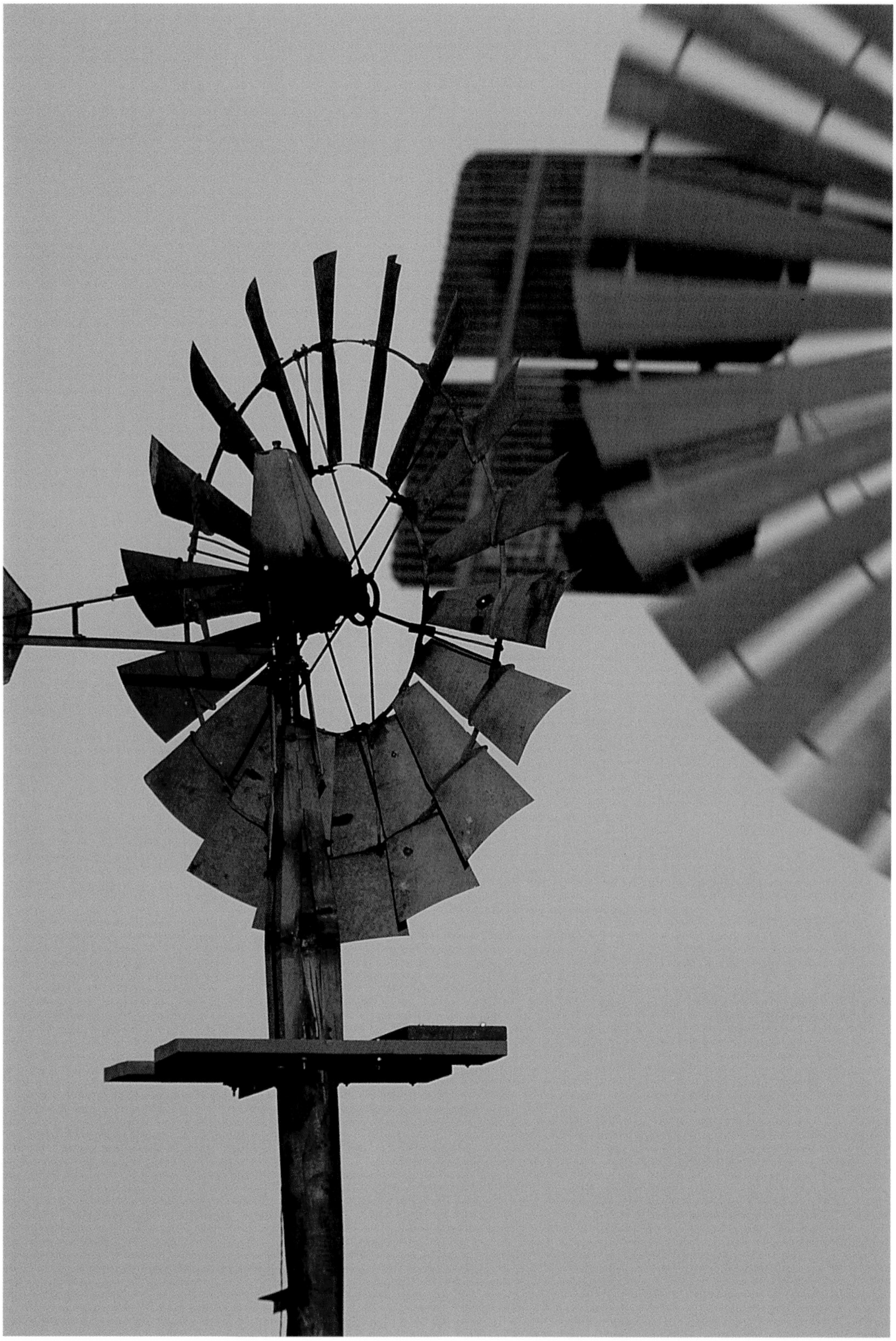

Foreword

IN ABOUT 1960 I first encountered Daniel Halladay in my junior high school history book. After seeing his name and reference to his invention of a windmill, I asked my father if we were related. Sure enough, he was one of my great-grandfather's brothers. I must admit that this made quite an impression on me. After all, seeing your family name in a textbook is pretty heady stuff for a young teenager. Daniel was right in there with George Washington, Robert E. Lee, Eli Whitney, and others.

The fact is, though, I had not fully grasped the nature of his invention. And for probably the next twenty years I was certain that Uncle Dan had invented "the" windmill. Not so. Some say that the Europeans had windmills as early as A.D. 1000. What he designed and patented in 1854 was a mill that could regulate itself to wind speed. His initial shop was in South Coventry, Connecticut. In 1863 the company, which had now become the U.S. Wind Engine & Pump Company, moved to Batavia, Illinois. Interestingly, the majority of the company's buildings still survive in that beautiful town adjacent to the Fox River.

In the late 1970s my father and I started a leisurely quest to learn more about Daniel Halladay and his windmill. At an antique show Dad found an 1890s catalog, which convinced us that the U.S. Wind Engine & Pump Co. manufactured the Halladay Standard mill and a lot of other things. Somewhere I located the name of the late M. I. Rasmussen, who taught a windmill maintenance and repair course at New Mexico State University. I called that gentleman, who was still living and teaching at the time, in search of details about the Halladay Windmill. Although he was a wealth of information about windmills in general, he didn't have a lot of data about the Halladay. Being somewhat disappointed, I asked Professor Rasmussen if he might know of an authority on windmills who could answer my

U.S. WIND ENGINE
AND
PUMP CO.
HALLADAY STANDARD
BATAVIA, ILL., U.S.A.
WIND MILLS, PUMPS, TANKS AND HAY TOOLS.
BRANCH HOUSES, KANSAS CITY, MO. OMAHA, NEB.
DEPOTS, FORT WORTH, TEX. BOSTON, MASS.

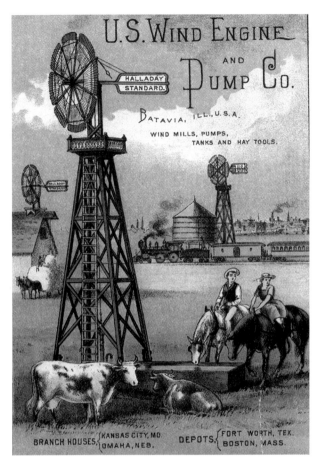

many questions. His response, which pretty much ended our visit, was, "Sir, I am the authority on windmills."

Taking him at his word, I reasoned that I might never find out much about the subject. To the rescue came T. Lindsay Baker and his book *A Field Guide to American Windmills.* That comprehensive book is an incredible source of information about the many windmill companies that existed during the 1800s and 1900s. Dr. Baker lives and works only a couple of hours from our home in Texas, and we have formed a friendship based on my many questions and his patient answers. He is directly responsible for my locating an unrestored Halladay Standard mill, the restoration of which will probably take the remainder of my natural life.

A Halladay Standard windmill pumping water for a home in Iowa, about 1900.

My search for windmill information has been, and remains, a delightful experience. We owe enormous gratitude to the pioneers of windmill collecting, such as the late Billie Wolfe and J. B. Buchanan. The annual International Windmillers' Trade Fair brings together wonderful people with a common interest in preserving the old mills. At these functions you can meet B. H. "Tex" Burdick, a centenarian who started erecting mills in the 1920s; Ken and Sharen O'Brock, whose vocation for thirty-five years has been selling and erecting mills; T. Lindsay Baker, who is so generous with his knowledge; and literally hundreds of other enthusiasts.

The pages of this book are filled with pictures of vintage windmills. As you can see, they are more than mechanical devices. Many of them are works of art. Through the vision of Billie Wolfe and the stewardship of Coy Harris, the American Wind Power Center has collected and restored over one hundred of these wonderful mills. They are preserving a portion of our heritage.

We have often heard that the Winchester won the West. There is probably some truth to that. But it was the self-regulating windmill that provided the water that allowed farmers and ranchers to settle the American plains.

STEVE HALLADAY
Austin, Texas
July 2003

Windmill Tales

ON A RANCH just west of Abernathy, Texas, in the late 1920s a rancher had bought a Twin Wheel windmill. The two wheels were unusual enough, but it was made even odder by its very short tower. The bottom of the wheels just cleared the ground. When folks drove out from town to watch this big windmill work, they heard the rancher explain the stubby tower: "It's short because there is a whole lot more wind down here on the ground. I can feel it on my face."

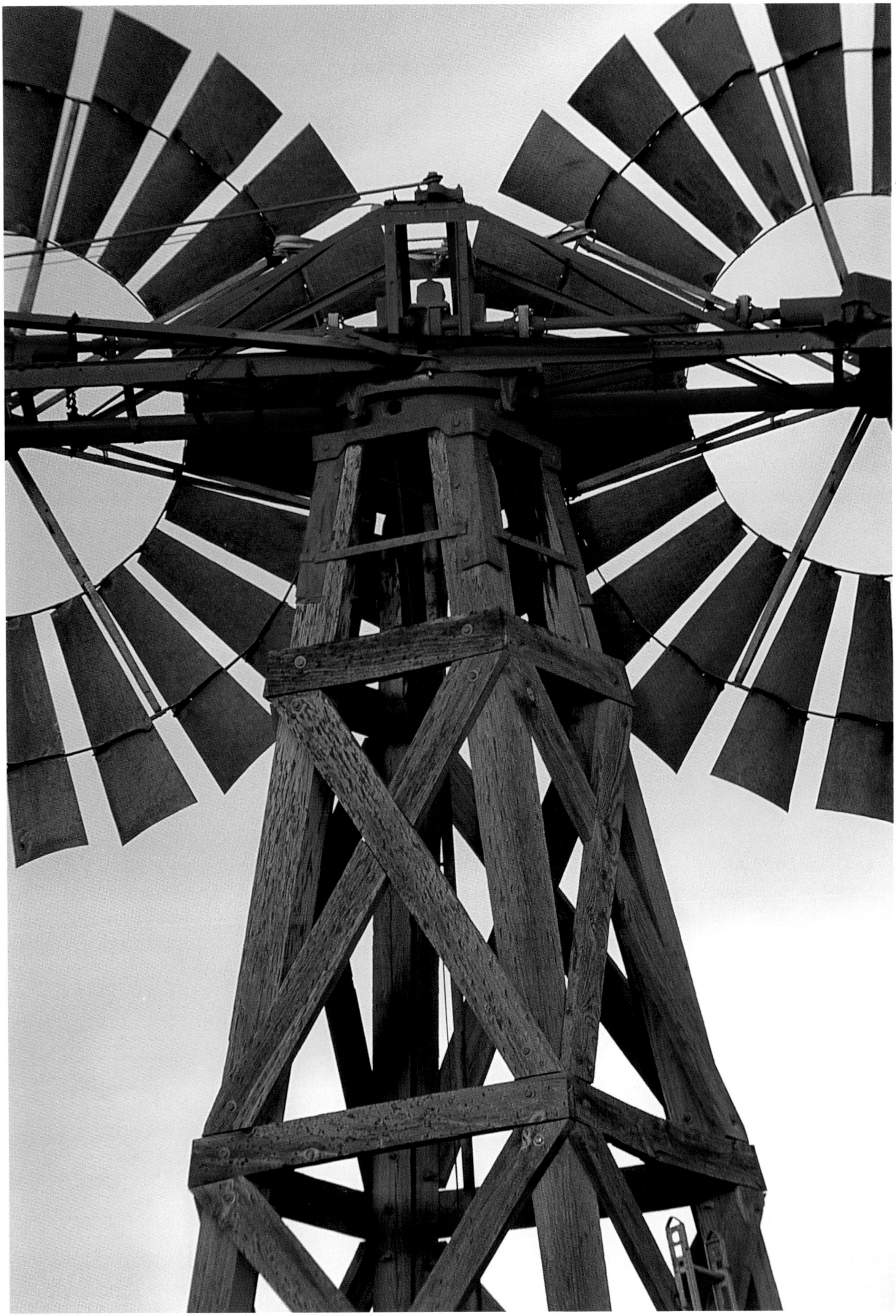

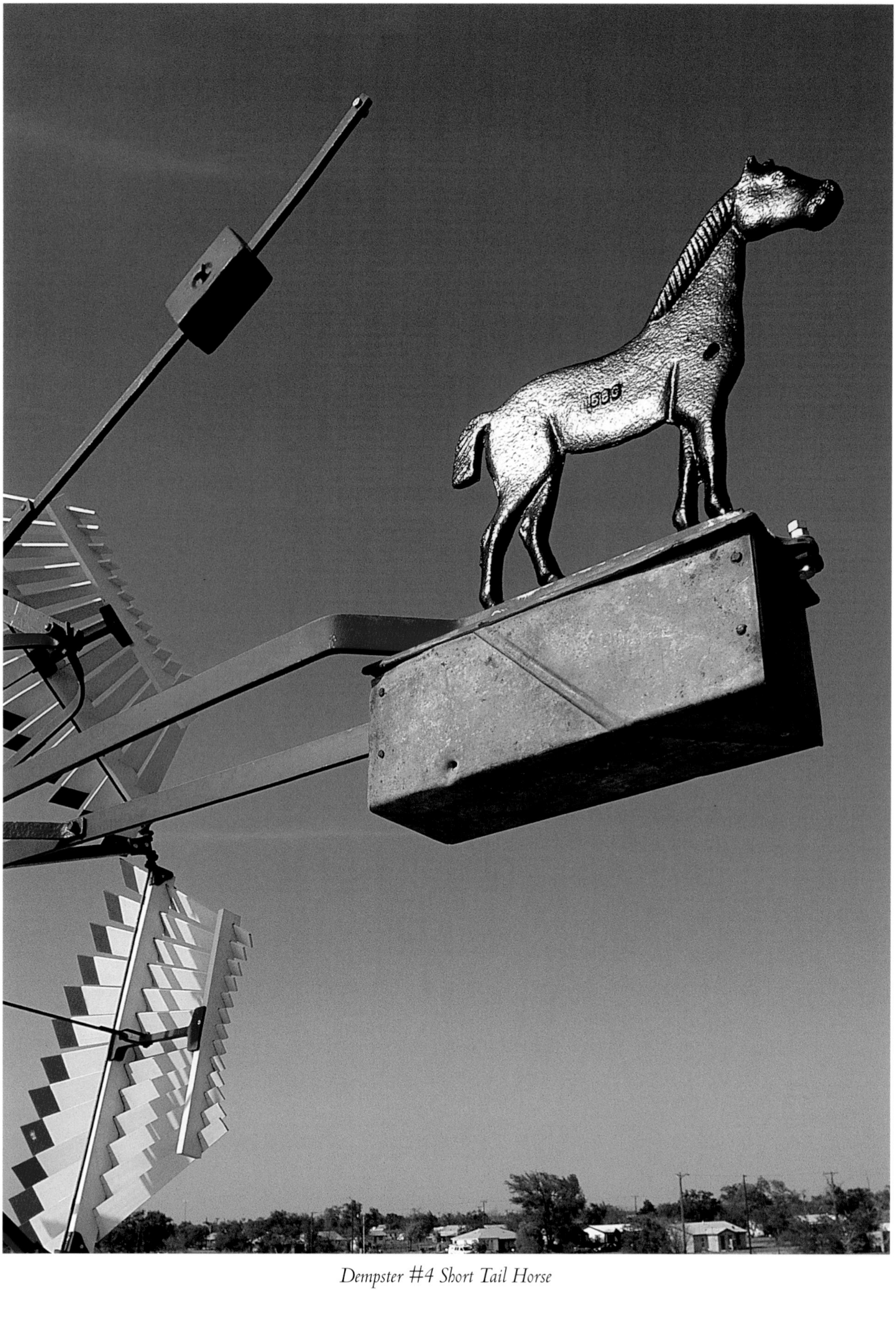

Dempster #4 Short Tail Horse

The Windmill Tales

ON THE PRAIRIES AND PLAINS OF NORTH AMERICA, water in a quantity needed for agriculture was hidden underground. The vast grasslands sustained great migrating herds of animals, which in turn sustained native Americans, but it was not until water could be brought to the surface that the plains could be cultivated and developed into a great bread basket for the growing nation. The portable, affordable, self-governing windmill forever changed the culture of this vast region.

The agricultural development of the plains is the story of the ingenuity, hardship, success, and sometimes failure of settlers as they applied a new technology in an environment with which they were barely familiar. The stories of these settlers and of their descendants often focus on the windmill, for that source of life-sustaining water often became the center of ranch and farm life.

A family's windmill stood on a metal tower usually twenty-seven feet tall. An eight-foot-diameter steel wheel was most common, and it pumped water from a well that was usually less than two hundred feet deep. It was good water, most of the time. Cool windmill water made for great swimming in the summer when the stock tanks were full. If the weather wasn't too bad, it was even fun to help pull the long lengths of wooden sucker rod when the windmill needed work. If your father let you climb up and help him, you could spit off the edge of the platform and see if you could hit those ants scurrying around down on the ground.

Over the years since 1993, when the American Wind Power Center was established, the museum's staff and volunteers have heard many visitors' recollections of living and working with windmills. It seems that almost every visitor has a windmill story, and their faces show the listener how much they enjoy repeating their tales. Early in the life of the Center, the staff and volunteers began writing down the stories as they heard them, and the result is a written record of this living oral history.

For many who grew up on farms and ranches in the West, climbing the windmill tower is a strong memory, and stories of climbing the tower are among the most often repeated at the Center. Without trees nearby, the windmill was the tallest object around. Children felt compelled to climb it with pockets full of rocks and flying objects to drop from the platform. Although Mother warned them not to, how could they resist? "I'll be safe," they'd say, "'cause I've seen Dad do it a hundred times." Climbing the windmill tower was different from other daily events. There was an element of risk in it, and that risk left impressions. In many cases an encounter with the windmill was one of life's first outdoor challenges.

Other visitors to the American Wind Power Center recall the creaking, metallic rhythm of the turning wheel on a windy day, the coolness of the clear water surging from deep in the ground and spilling into a wide tank, the silhouette of the wheel against a blue sky, or the constant change of clouds passing overhead. Still others remember the difficulties of keeping the windmill operating smoothly, the attention it needed, and even accidents involved with the tower or its machinery—most not serious, but some life-threatening, usually when a spinning wheel or swinging tail would wipe a person clean off the tower.

Almost all the stories are of pleasant memories. Most are brief, yet together they give a sense of what it was like growing up on a farm or ranch before electricity made life somewhat easier. When guests to the museum spot a windmill that reminds them of the farm where they grew up, they begin talking: "Listen, when I was growing up, our windmill . . . ," or, "Do you know what our momma did after we climbed the tower? . . ."

Many windmills are gone now, fallen in disrepair when the children left home or after the parents passed away and no one was around to keep the wheels turning. When you see a windmill from the road, you wish it was a little closer, because you can just hear echoes of the whistle of the wind through its vanes. Those are good memories worth retelling as a windmill tale.

WYMAN MEINZER'S photography in *Windmill Tales* gives the reader a chance to see a portion of his work that only occasionally appears in his other books. Those books might have a glimpse of a windmill, but here the windmill is the object of his camera.

There is history in these photographs. The windmill meant life to thousands of people who had to struggle to live in areas where there was little surface water. Today, windmills bring back warm memories to those who lived with them and had to work on them and who appreciated the windmill as their only source of water.

Many windmills are derelict now, stopped in their work by changes in technology, the passing of their owners, the availability of electricity, and sometimes unknowing neglect. These windmills might have a blade or two missing, a bent, bullet-holed tail, or birds' nests in the machinery, or they may have taken that deadly plunge from the top of the tower. Wyman's camera has captured them as well as those that are still in use and those that have been preserved by the American Wind Power Center, and in these photographs he has translated the appealing character of all windmills.

Wyman and his wife, Sylinda, photographing windmills.

PHOTO BY COY F. HARRIS

"MY BROTHER AND I had to do all the maintenance
on our family windmill. One night before we were supposed to oil the mill, a big rainstorm came up and flooded the entire pasture. The next morning we had to walk
to the windmill through mud that was knee deep. We had
always been told to tie the wheel off when we worked on
it, but this time we didn't. While we were working a gust
of wind came up and whipped my brother clean off the
platform. I looked down expecting to see him dead, but
there he was standing up just like he landed,
in mud up past his knees."

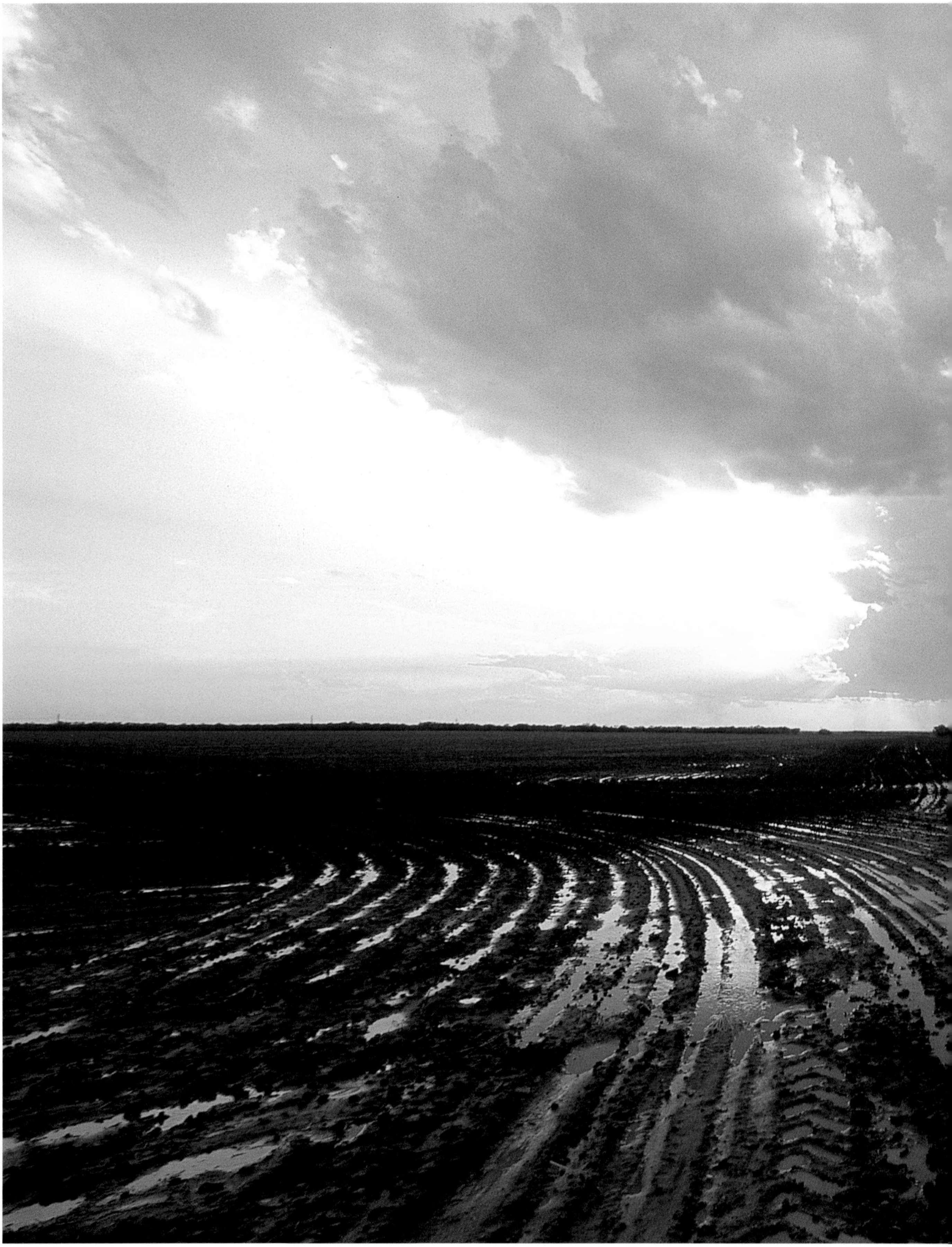

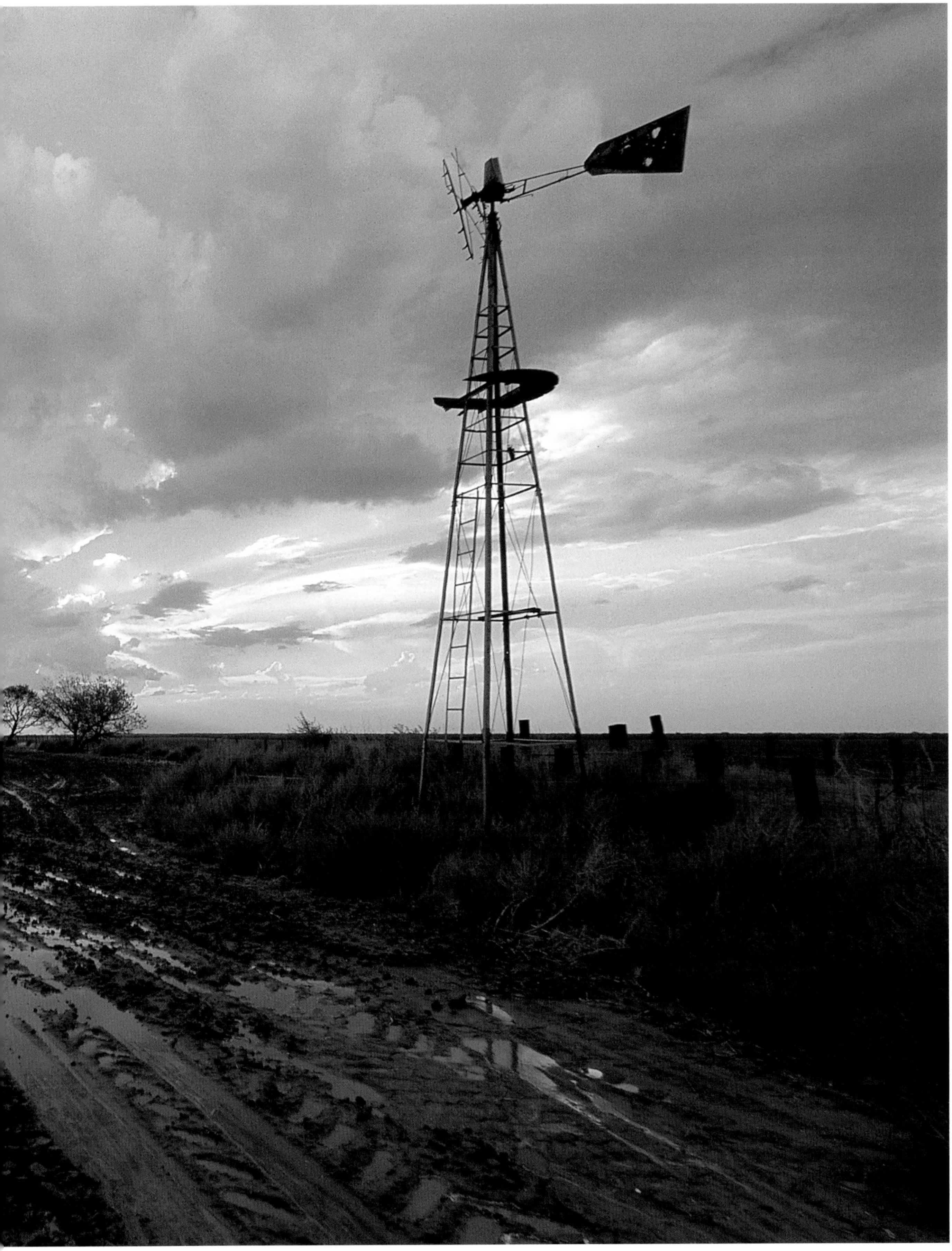

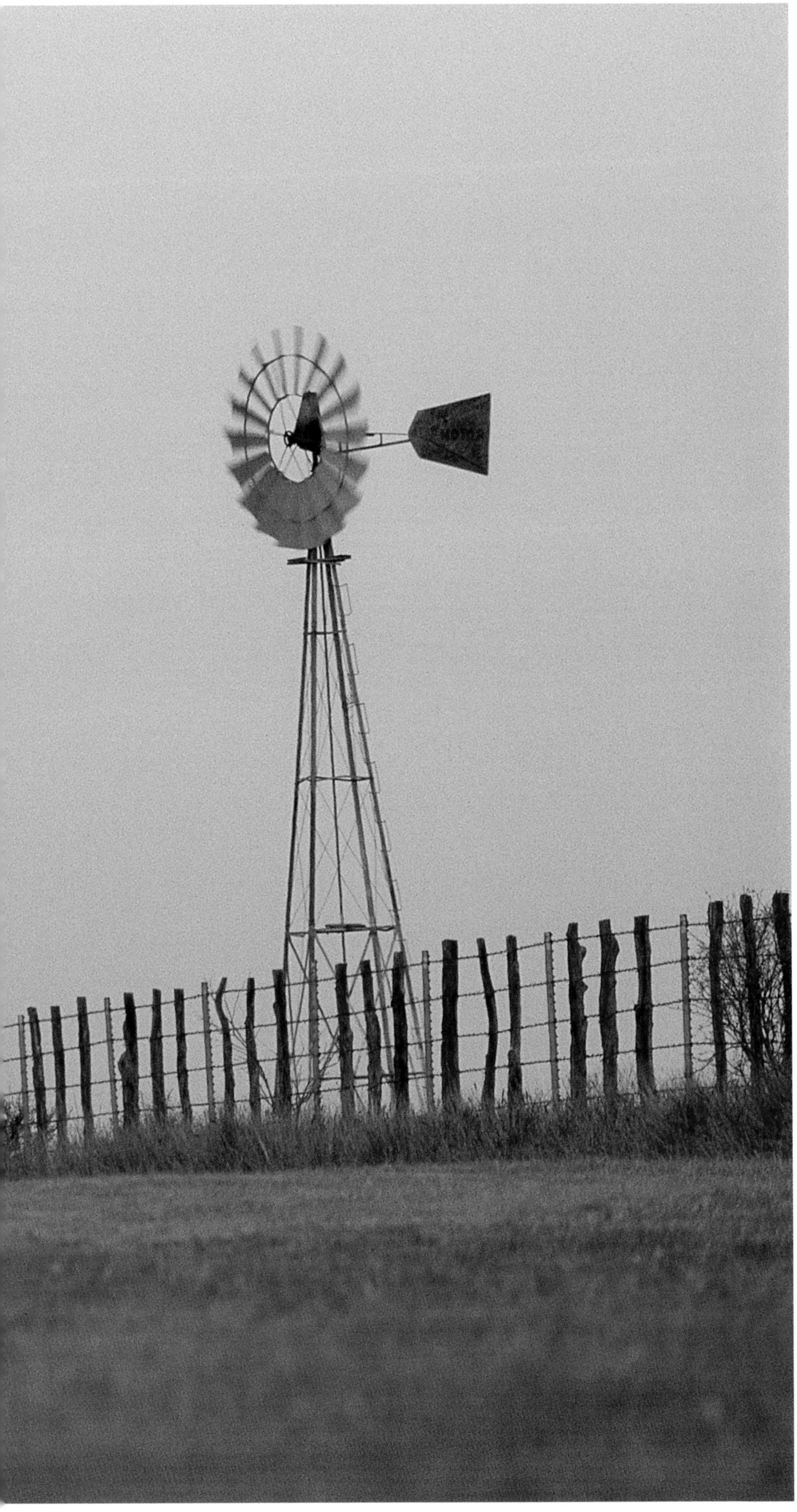

CAROL WAS ABOUT EIGHT YEARS OLD when she lived on a cattle ranch. In the spring when the wind blew hard, all of the fences would have tumbleweeds stacked thick against them. One particular cow would go and hide in the tumbleweeds against the fence when she was ready to calve. It was Carol's job to climb the windmill and scan the fenceline looking for the cow and her calf. She said she loved the job because that was the only time that she was allowed to climb the windmill tower.

"**M**Y UNCLE WORKED THE WINDMILLS on ranches in New Mexico, and he had forty-seven that he kept pumping. On one trip he took me with him to replace the wooden sucker rod on one of those mills. We replaced all the rod that screws together, but the final section bolts to the steel pump rod. When we started to bolt these together, my uncle realized that he had not brought a brace and bit and couldn't drill any holes. So he got his .30-30 rifle, stood off a piece, and shot two holes in the wooden rod. Then we could bolt the final sections together."

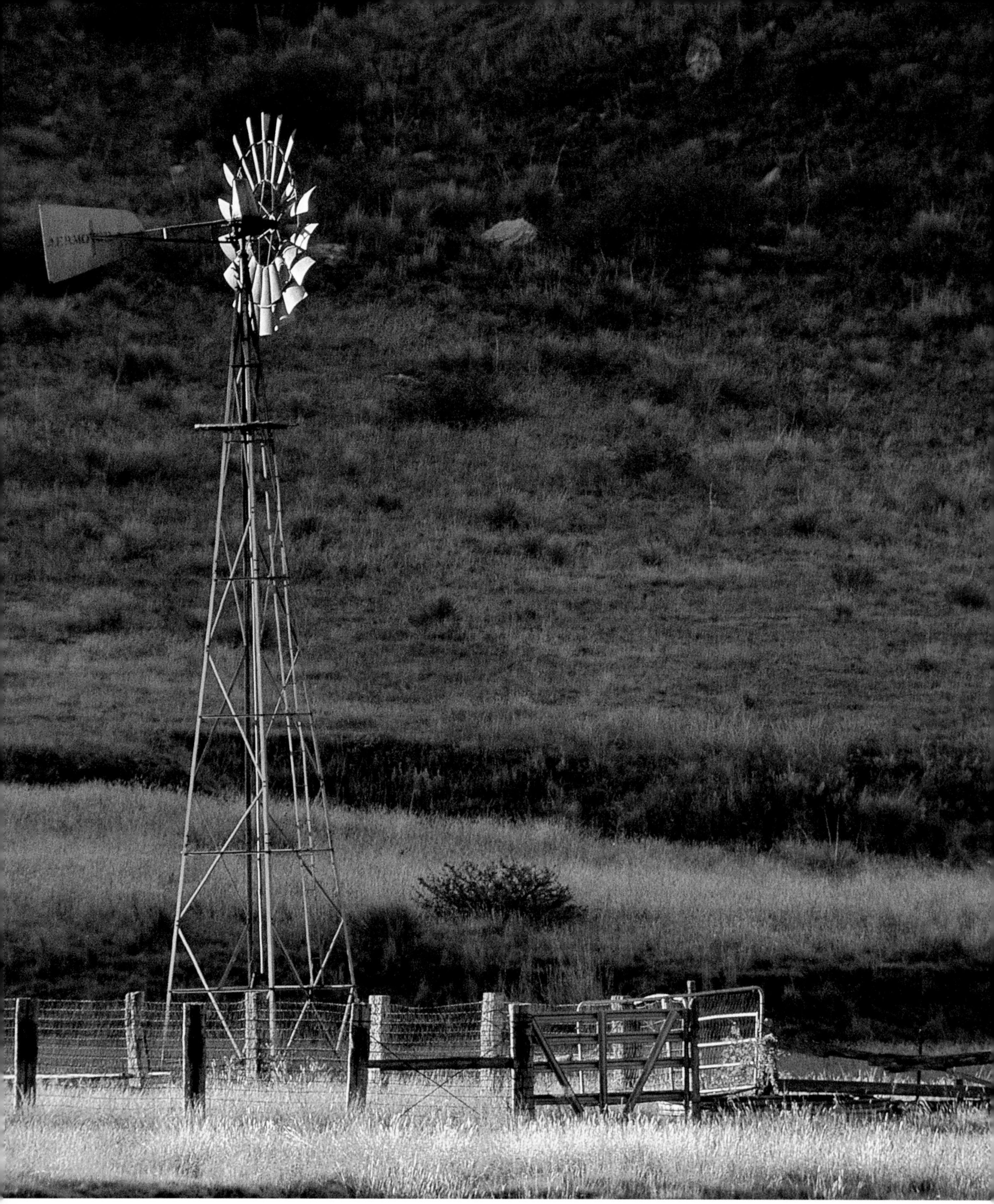

A COUPLE LIVED ON A RANCH that was several miles outside Roswell, New Mexico. On the road to their house was another ranch that had some of its windmills next to the road. One evening after dark they were coming home and noticed a strange light in the sky. As they slowed down to get a better look, they realized the light was coming from the top of their neighbor's windmill tower. When they stopped and looked up, the neighbor called down, "It's just me. I work on this thing at night because I'm afraid of heights."

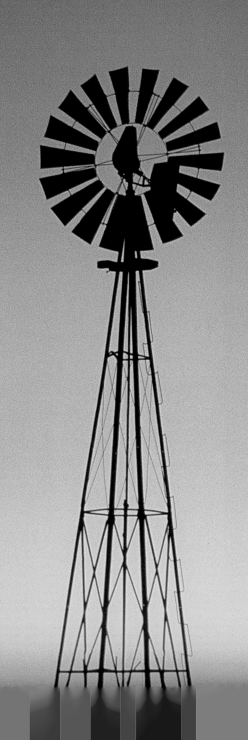

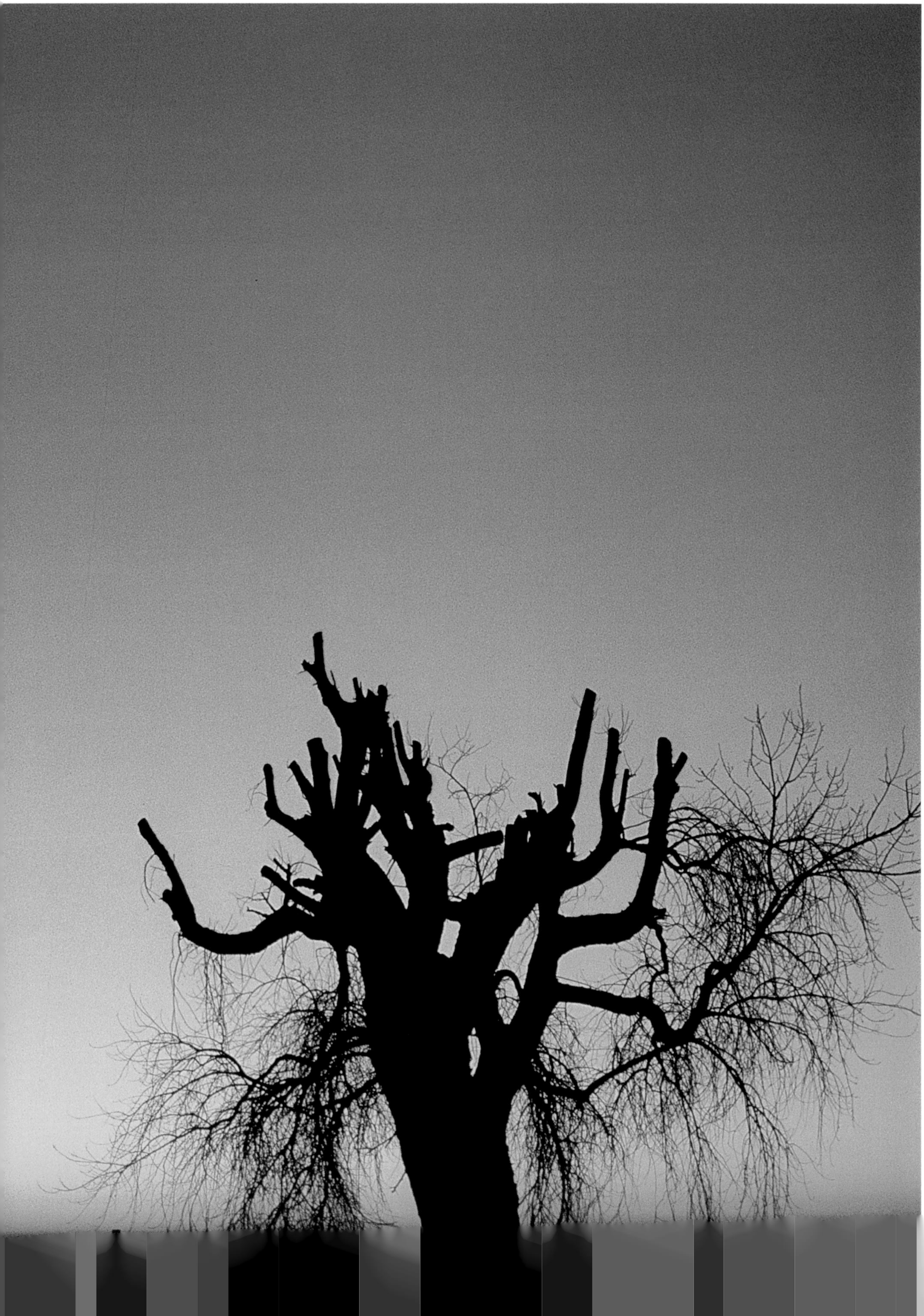

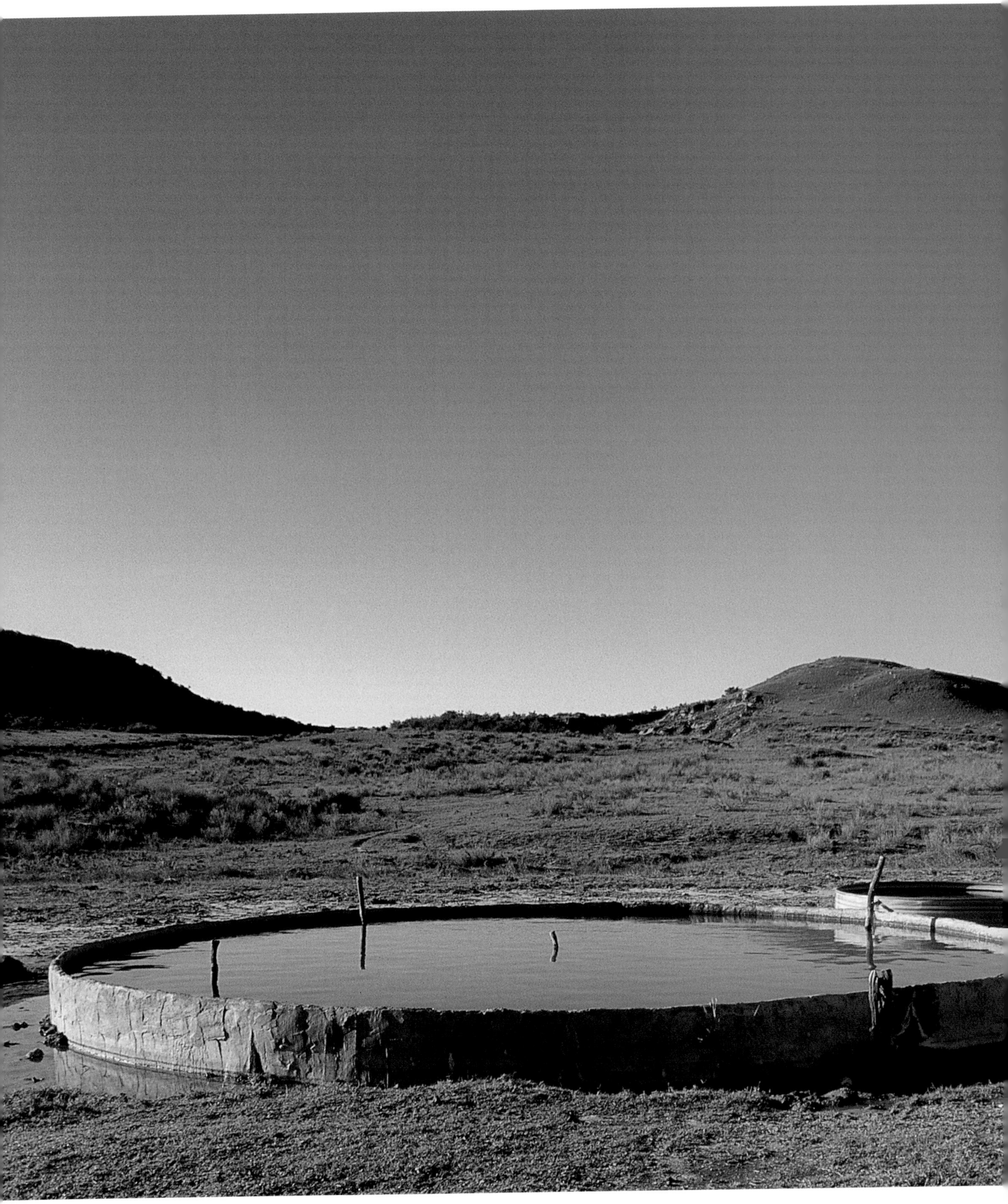

CHERYL WAS GIVEN A WONDERFUL Superman cape. She thought that if she "believed real hard" she would be able to fly while she was wearing it. One afternoon she decided to fly off the windmill tower that stood behind the house. After looking down at how far it was to the ground, she decided that just in case, she had better fly out over the water tank. She said that she must have done one of the biggest belly flops of all times, from twenty feet up.

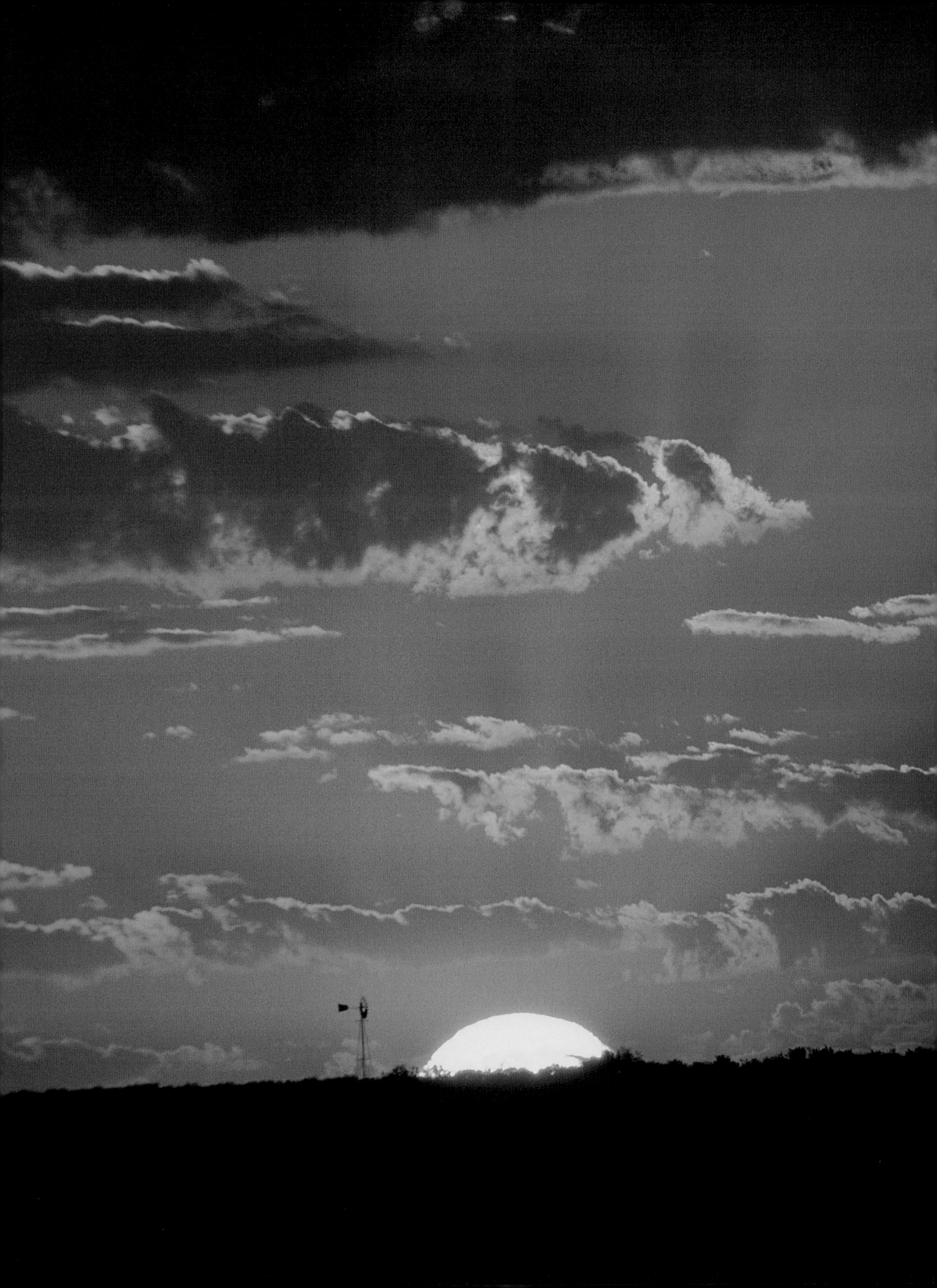

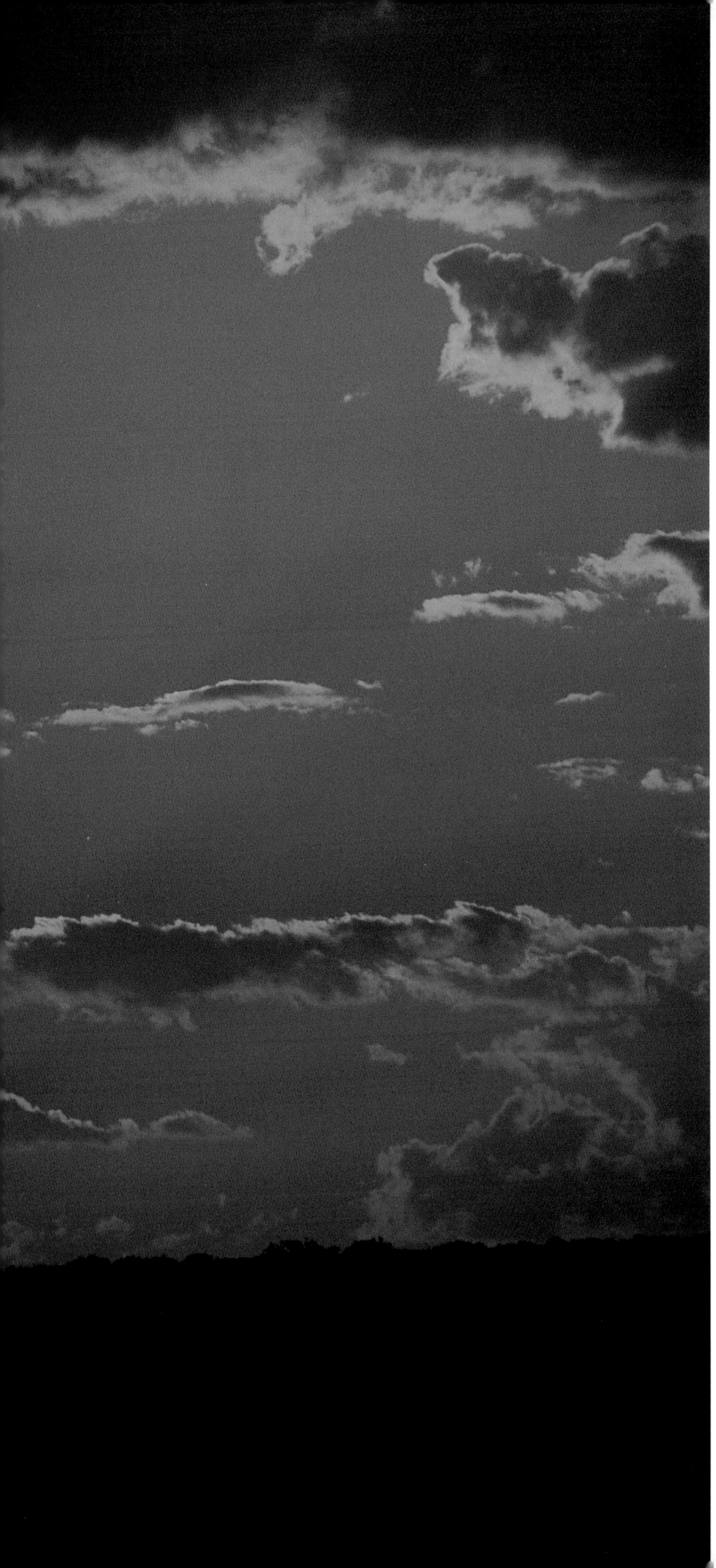

THE WINDMILL WAS USUALLY the tallest structure on the ranch. In flat country it was the only landmark for miles around.

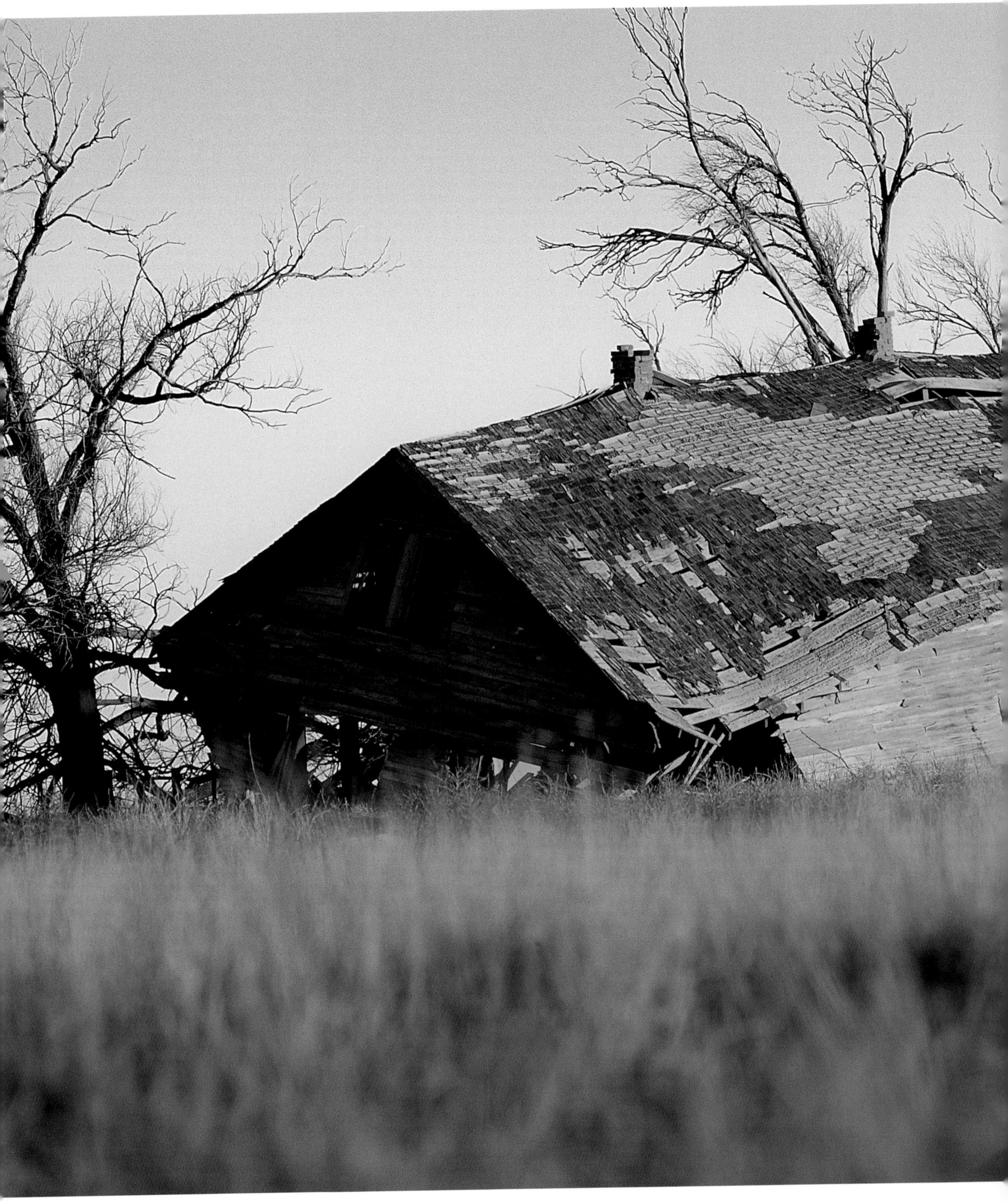

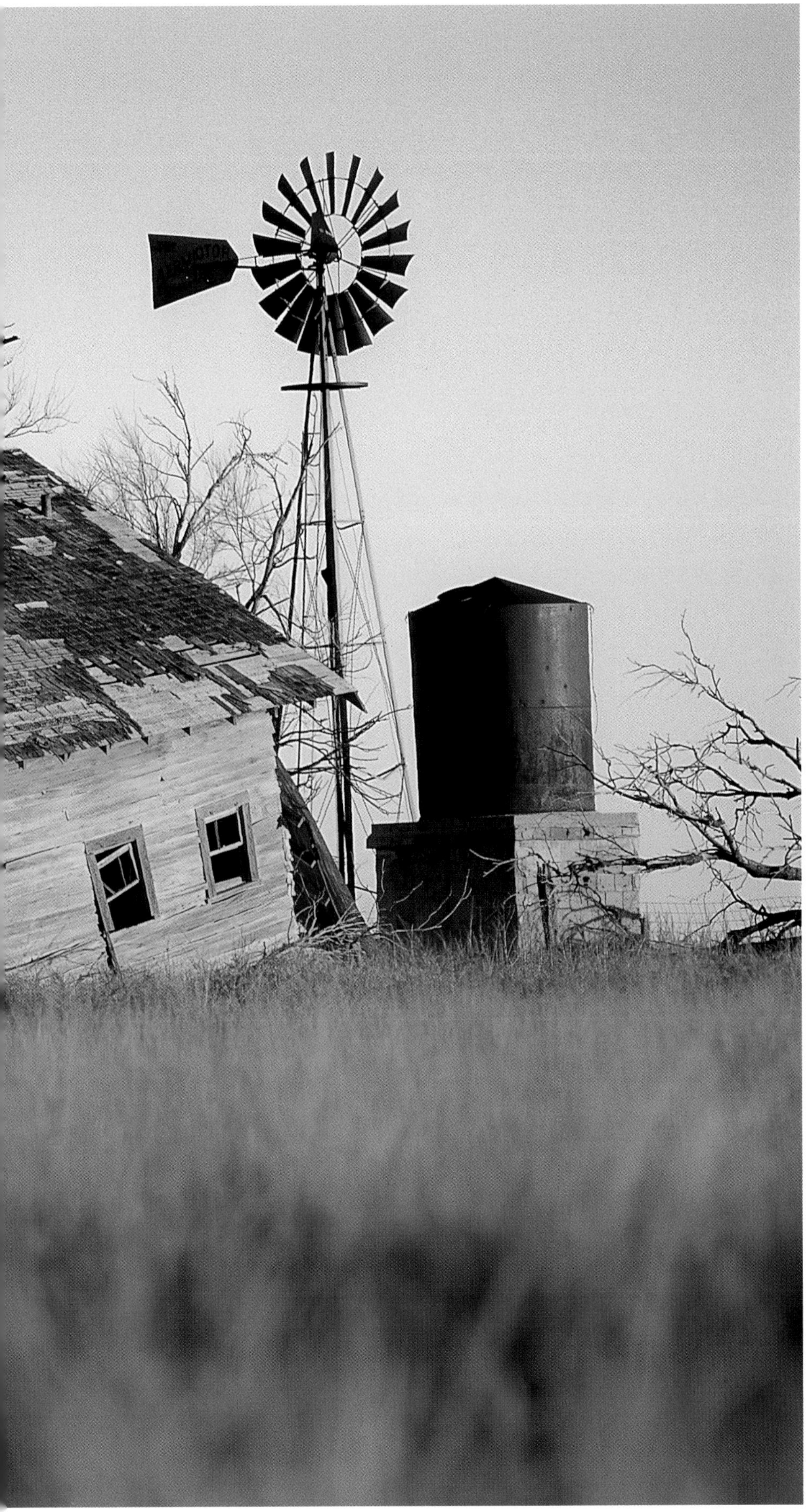

"My great aunt lived by herself on a small ranch in New Mexico. She was getting pretty old but could still do chores and liked to walk from the porch to the windmill. She had an old dog and a young cat for company, and they took care of each other. The last time I saw her she was standing by the windmill holding the cat. She was smiling and gave me a wave as I walked back to the car. I remember her smile while she stroked that cat and the dog lapped the windmill water. The only sound that day was from the windmill, a creaking sound that a windmill makes turning in the summer breeze. It is a memory I deeply cherish and is why I love windmills so much."

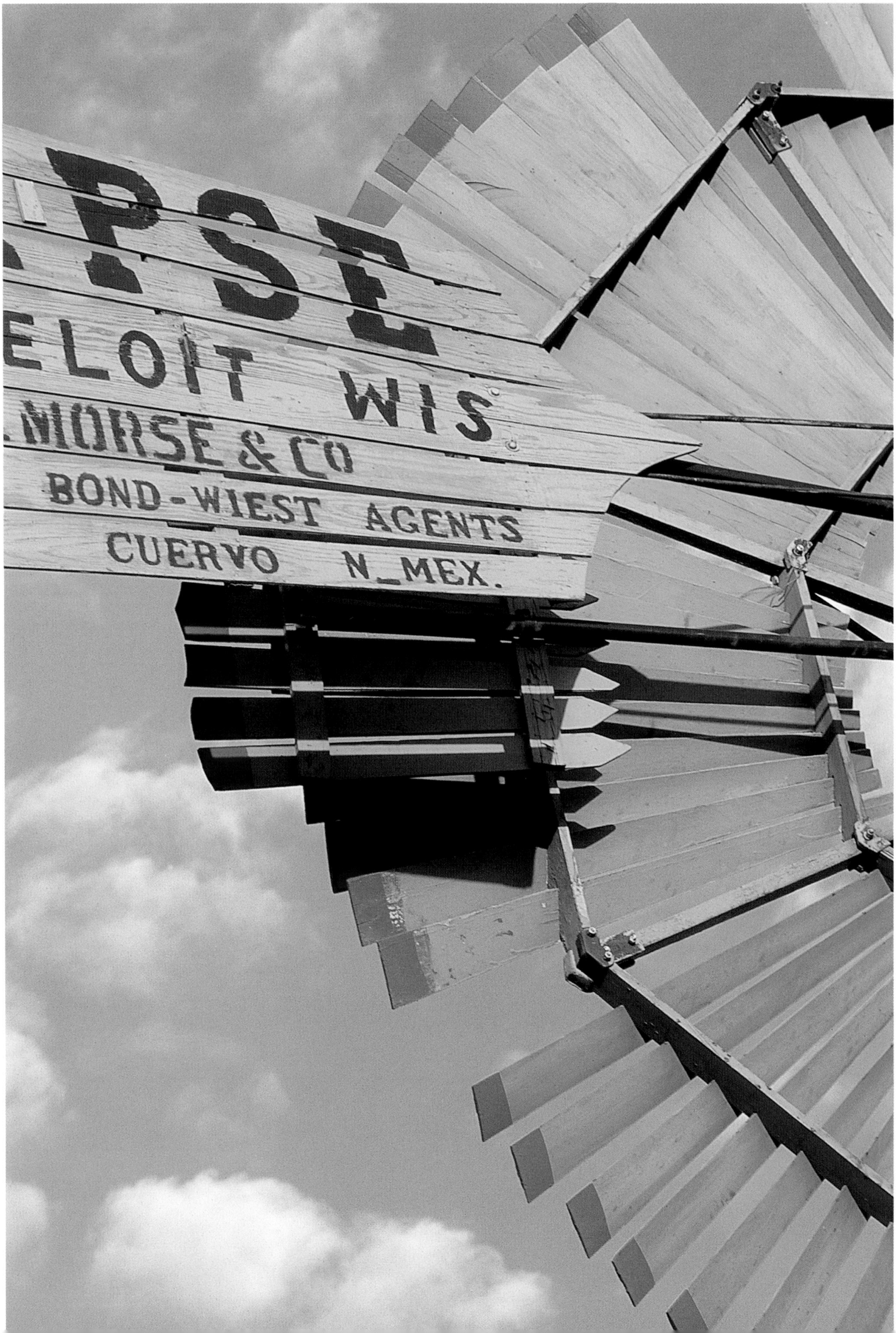
RSE
ELOIT WIS
MORSE & CO
BOND-WIEST AGENTS
CUERVO N_MEX.

THERE WAS AN OLD WOODEN Eclipse windmill behind our house that wasn't used anymore because it had been replaced with an electric pump. My mom had tied one end of the clothes line to a leg of that windmill, and the other was tied off to a post that stood near the back fence. One day a friend of mine was over playing, and we decided to untie the tower end of the line and retie it on the tail of the windmill. Mom had just hung several white shirts out to dry, and we thought they would make a great signal flag. I climbed up the tower and tied the line on the tail, and then we started playing captains on ships. A wind came up and twisted the tail around, which pulled on the line so hard it jerked the other end from the pole and pulled the shirts right through the dirt. Boy, did we get in trouble, and then my mom made us wash out the shirts."

A MUSEUM VISITOR SAID that one of the things he and his brother liked to do on warm days was to lie on the ground by the windmill and watch the sky. One day while gazing at the clouds the wind suddenly picked up, and the clouds began moving rapidly across the sky. He and his brother jumped up and ran into the house yelling, "The windmill is falling, the windmill is falling!" After a bit they realized it was just an illusion.

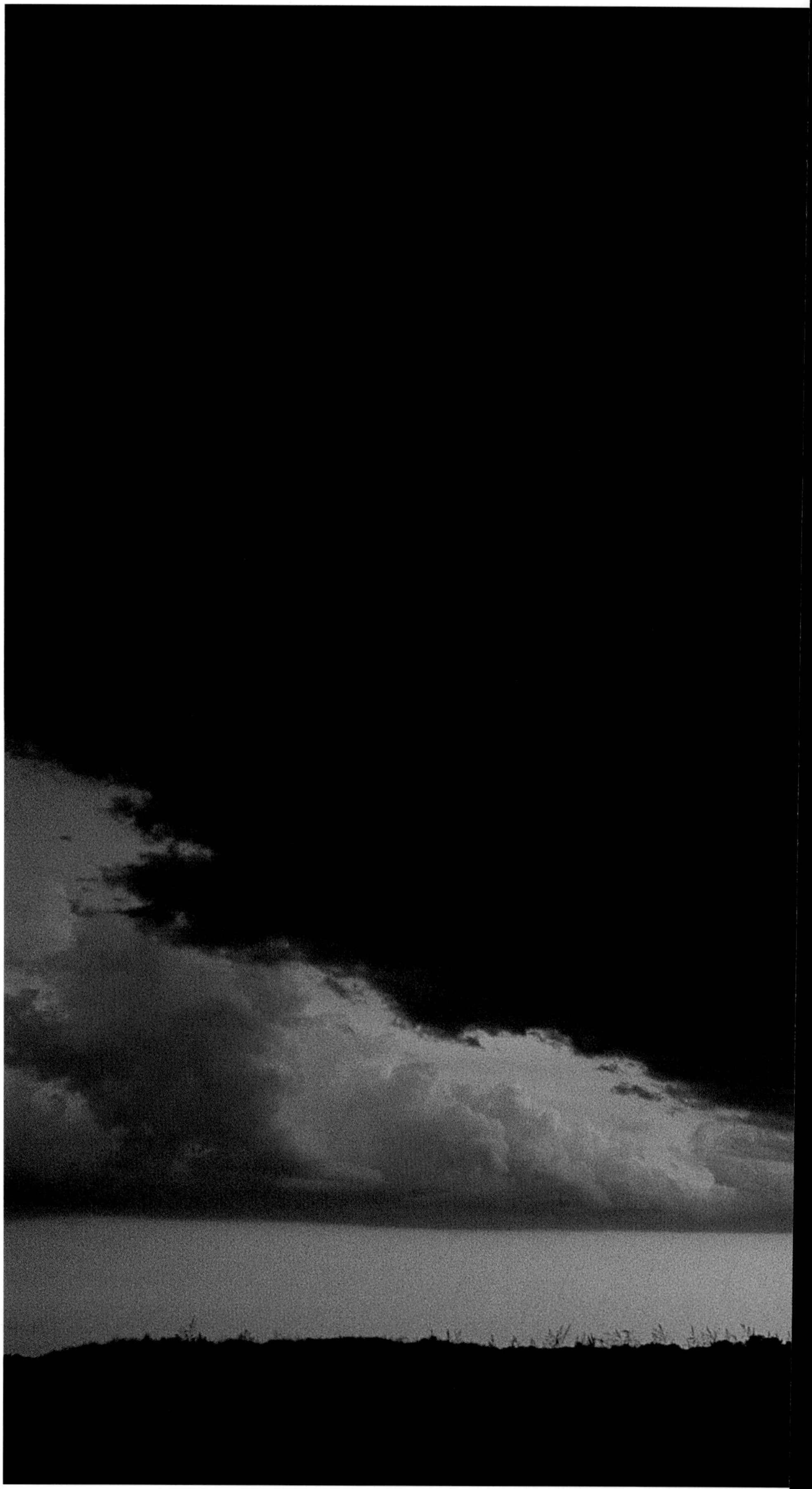

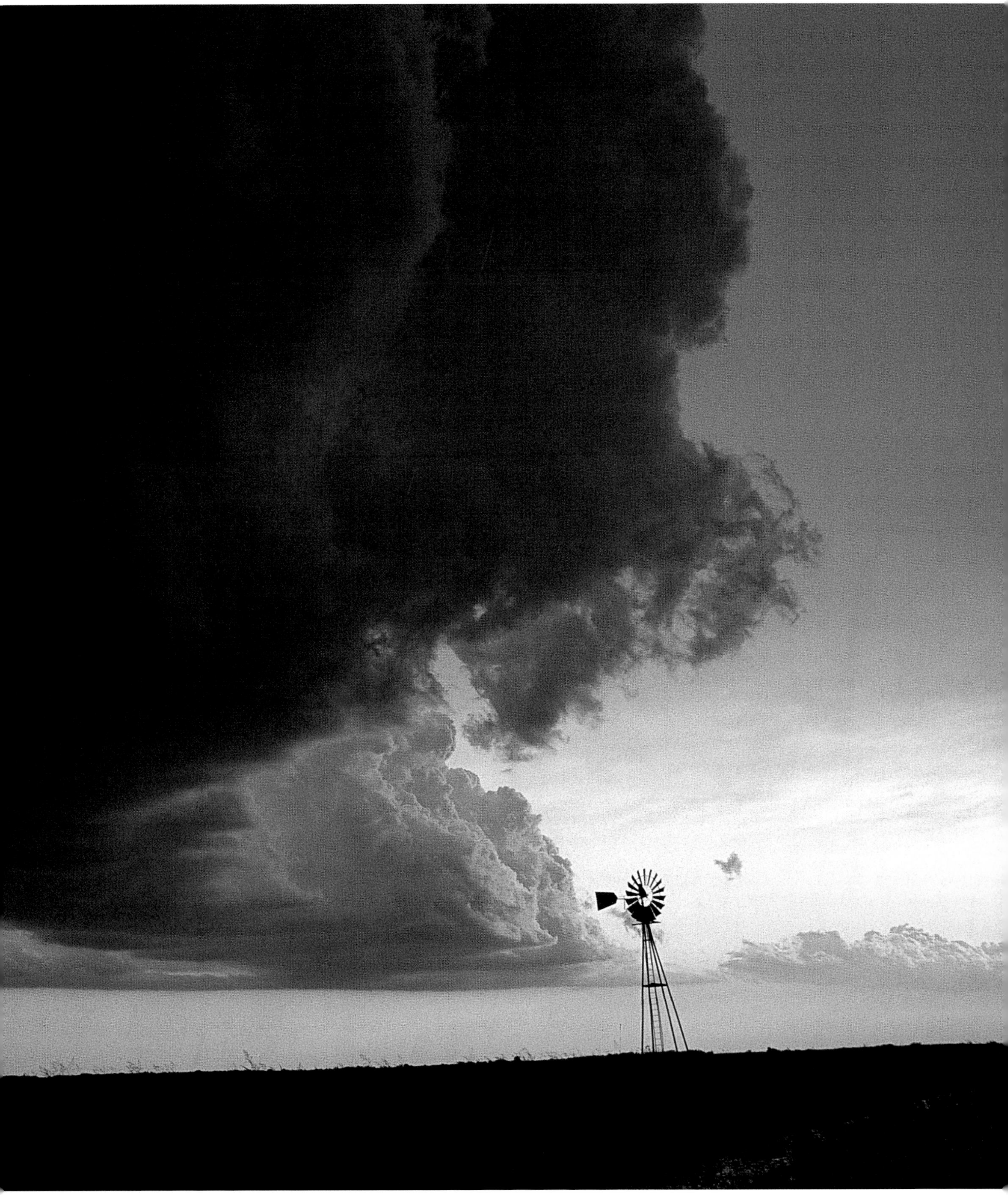

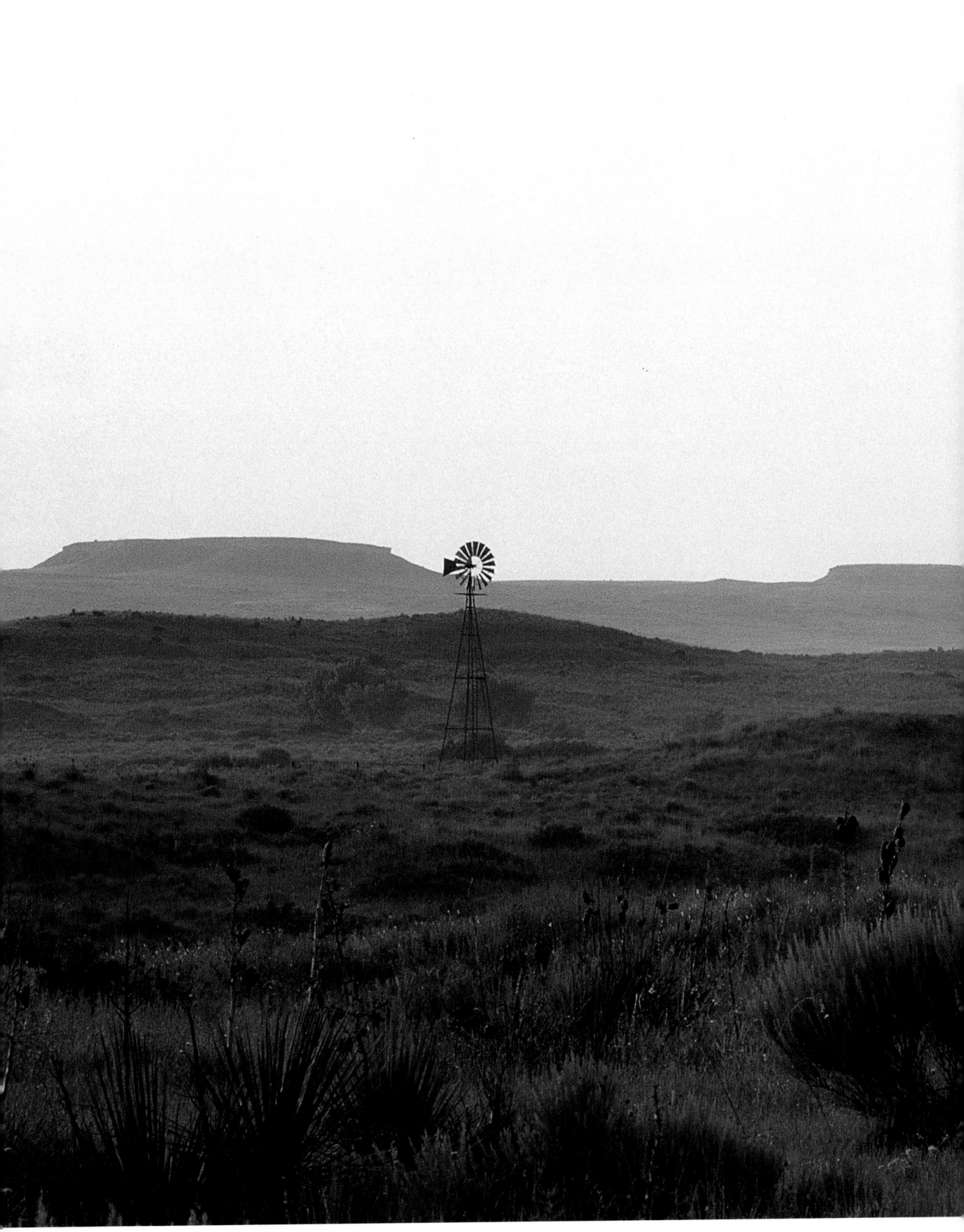

A GROUP OF MEN WERE standing around talking and noticed smoke off in the distance. One of them went to the windmill nearby and climbed to the top to see what was on fire. When he got to the platform he looked out over the countryside and yelled down, "Well, it's not my house!" He looked in the direction of the fire again, forgot where he was, and promptly walked right off the platform.

ONE FINELY DRESSED LADY said when she was a child in South Texas she would go out to the pasture to turn on the windmill. As soon as she neared the windmill, the cows would see her and start running toward her. She would begin screaming, "The cows are chasing me, the cows are chasing me." Actually, they knew the windmill was going to be turned on and that they would get some water. She said she spent a fair amount of time on the tower each time she had to start that windmill until all the cows got a long drink.

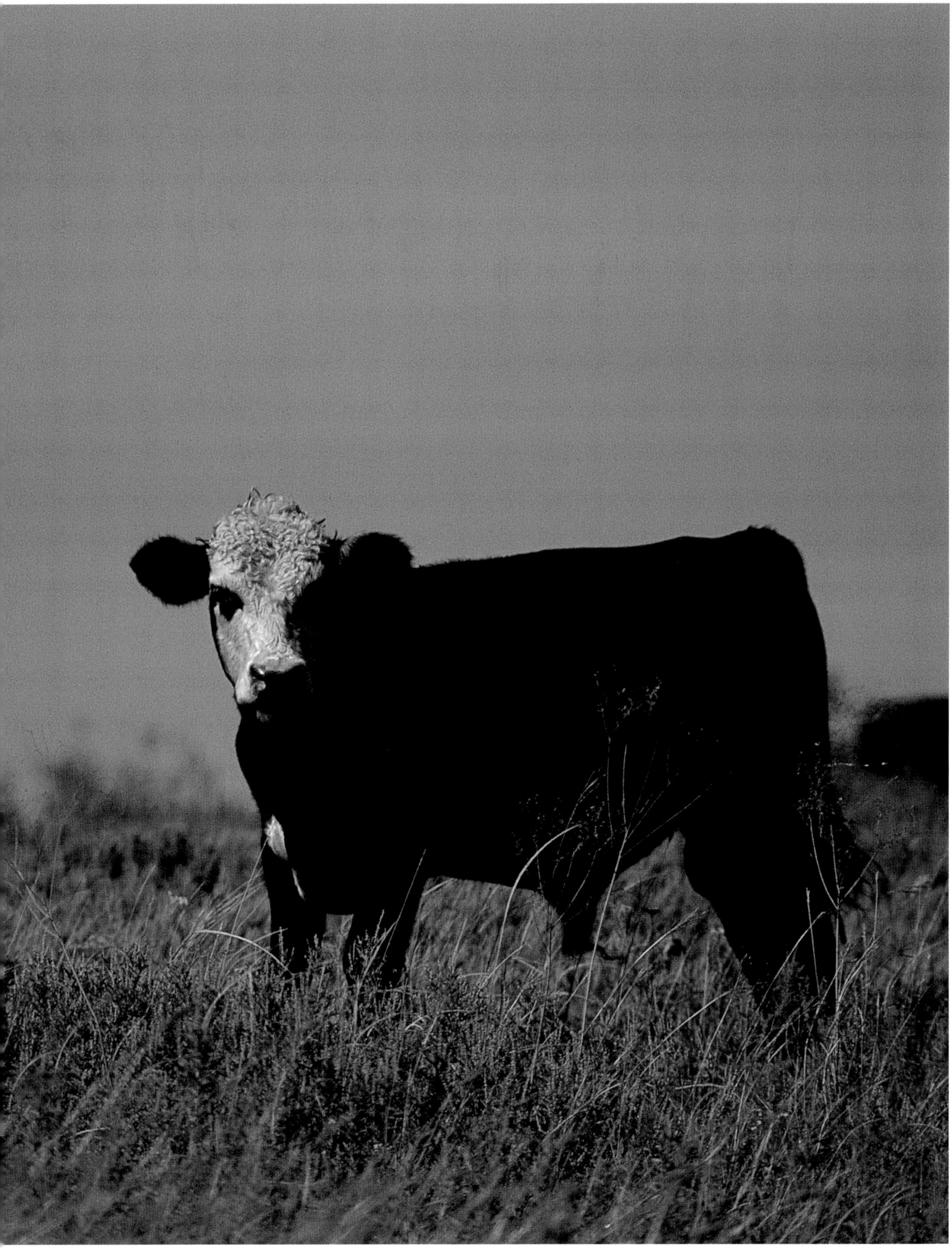

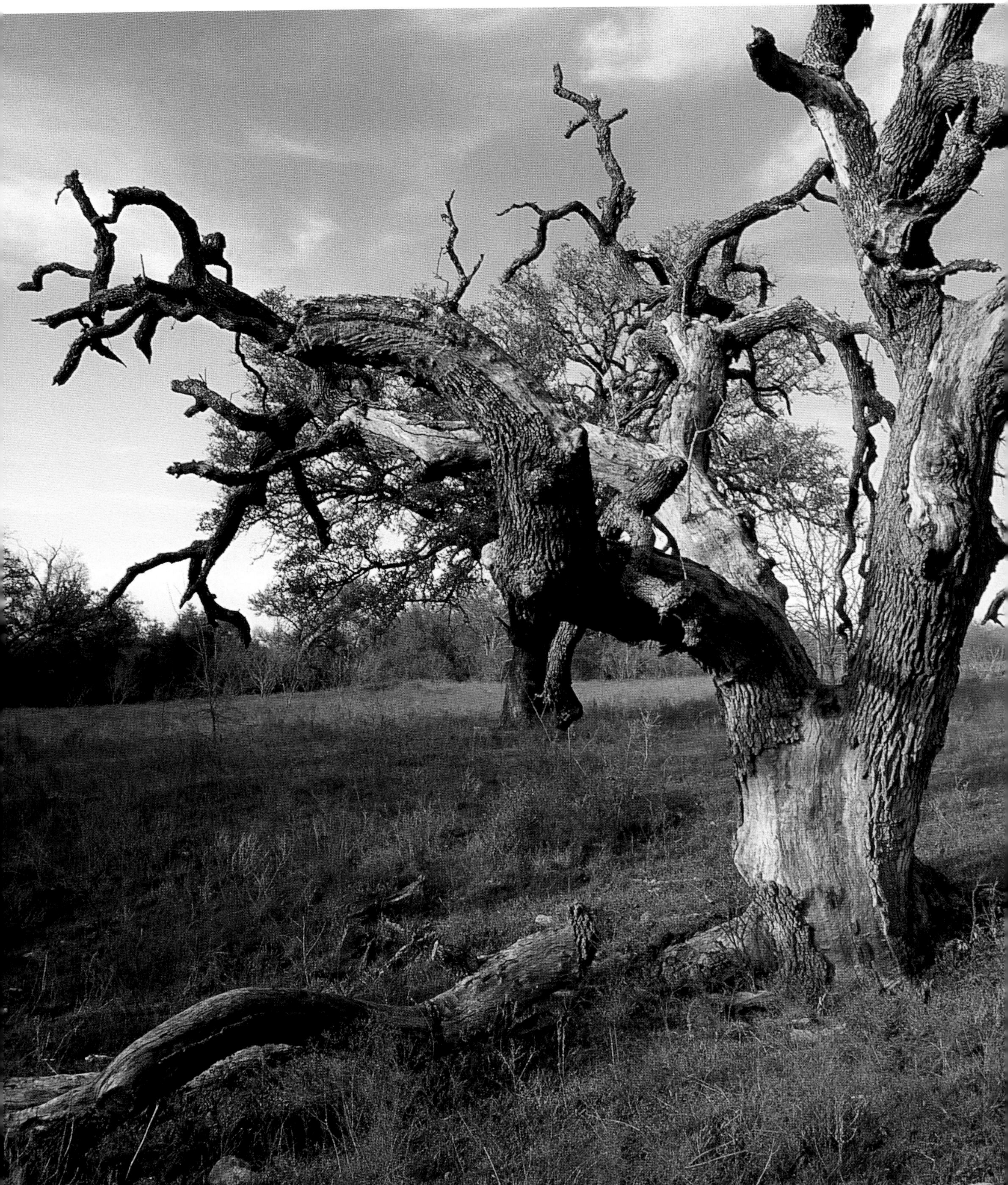

IN 1924 THE KAHLICK FAMILY moved to Slaton, Texas, where they have been farming cotton to this day. Mr. Kahlick said that he could never get a wrist watch to run correctly, so he never wore one. Mrs. Kahlick got tired of him not knowing when dinner and supper were ready, so she would go out to the windmill standing near the house and turn it off. When he looked up and saw that the mill was not running, he knew it was time to come in and eat. He said that is the first he ever heard of telling time by a windmill.

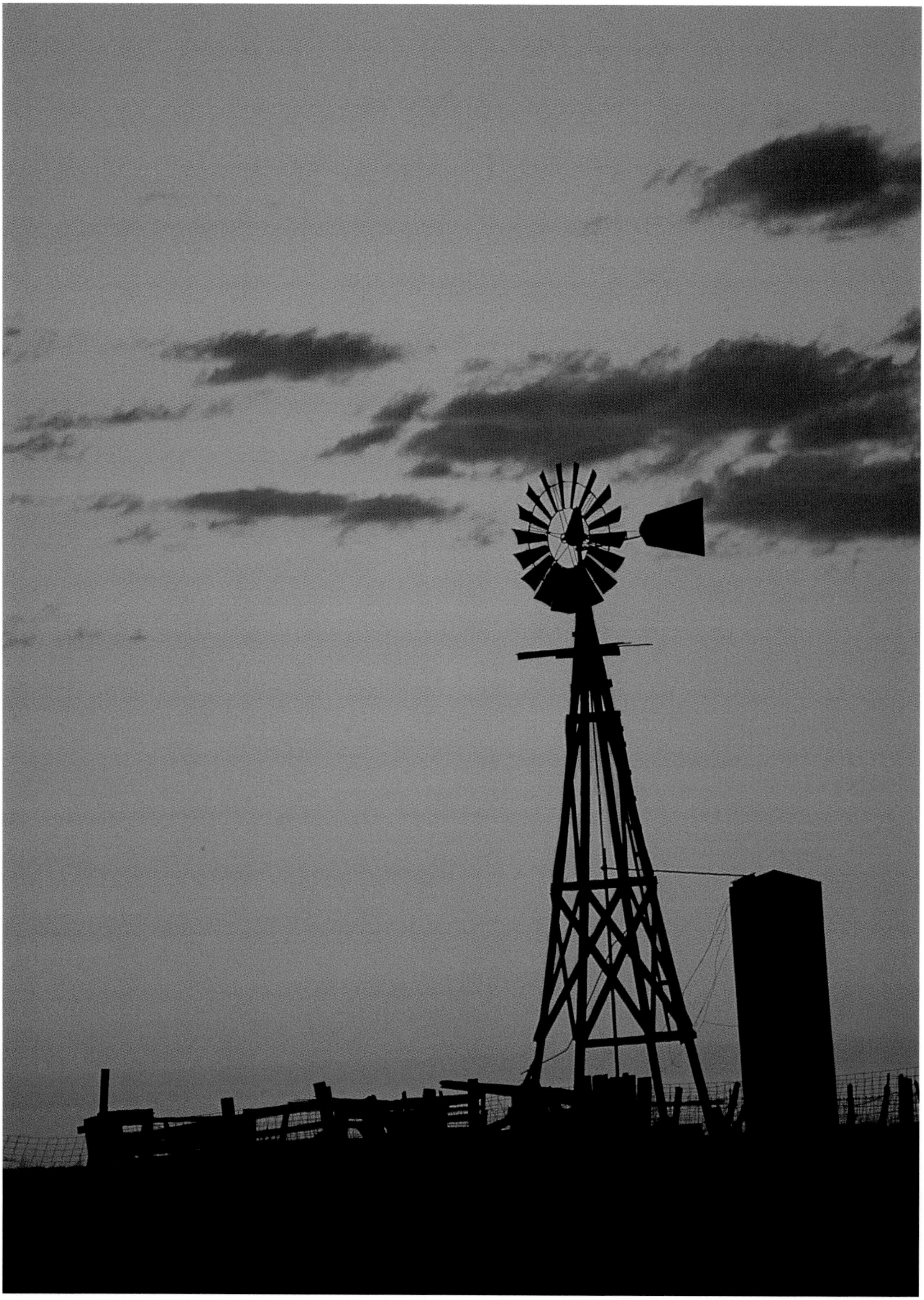

ONE DELIGHTFUL VISITOR said he grew up in the city and had never lived on a farm. He got engaged to a rural girl and was asked to visit his fiancée's farm to meet her family. They were all ready for the city boy. Her father gave him a bucket of grease and told him to climb the windmill tower, "'cause that windmill needed greasing." Not knowing that the windmill was supposed to be turned off before such work, he climbed the tower ladder, struggling with the bucket. Just as he put his head through the platform, the windmill whirled around, and the spinning wheel almost knocked him off. He looked down in desperation only to see everyone on the ground laughing. He climbed back down and told them he would never climb another windmill again. He did marry the daughter, though.

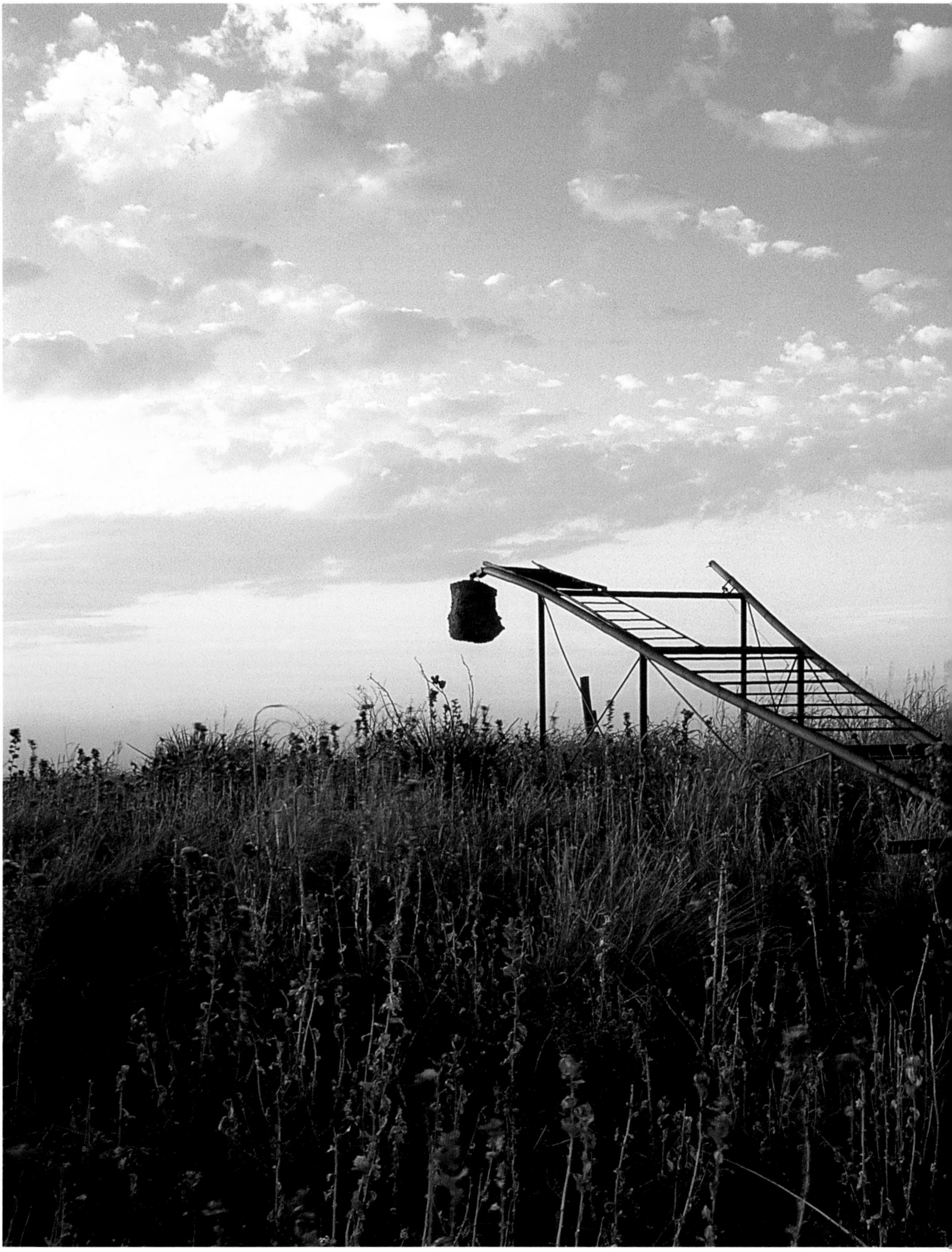

Just Plain Wore Out

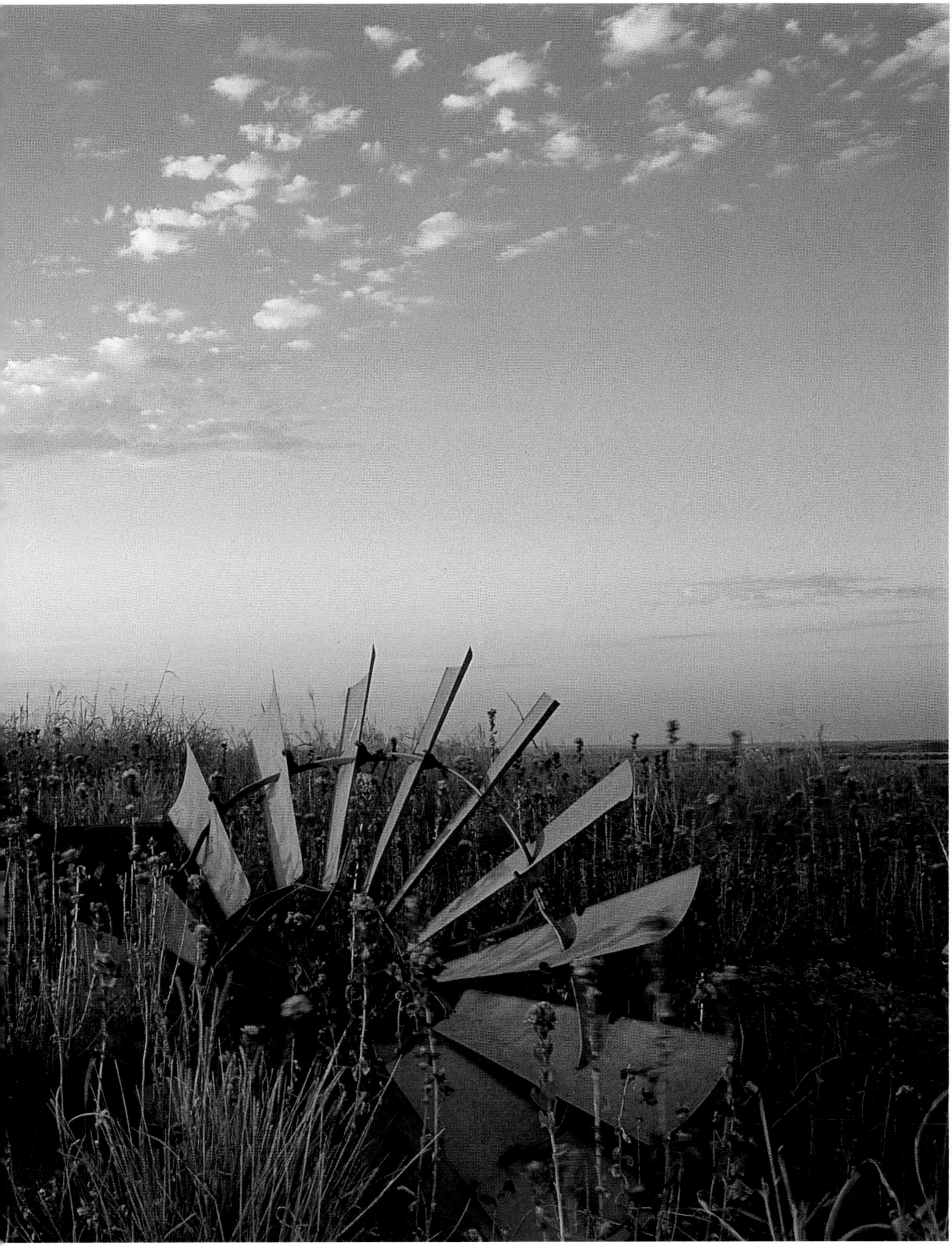

WHEN JOE WAS ABOUT EIGHT his dad erected a windmill on a hill a quarter mile from the house. Every evening about dusk it was Joe's job to walk down a long line of lilac bushes to the windmill and turn it off. He was afraid of the dark and usually asked if someone else would go to the mill, but they never did. Finally, one night his dad told him that he could take the kerosene lantern with him. As he was coming back with his light shining down the path, a "ghost" (his brother with a sheet over him) jumped out of the bushes and yelled, "I've come to get you little boy!" Joe swung the lantern and hit the ghost in the head as hard as he could, then ran to the house. When he walked in he had only the bail of the lantern in his hand, but announced, "I'm not afraid of the dark anymore. I even ran into a ghost." His dad noticed the bail of the lantern and said, "Oh my gawsh!" and ran out of the house. He came back dragging Joe's older brother, woozy from being knocked in the head. All that night and the next day Joe kept saying how proud he was of himself for outsmarting that ghost. It didn't sink in until a couple of days later that the windmill ghost had been his brother.

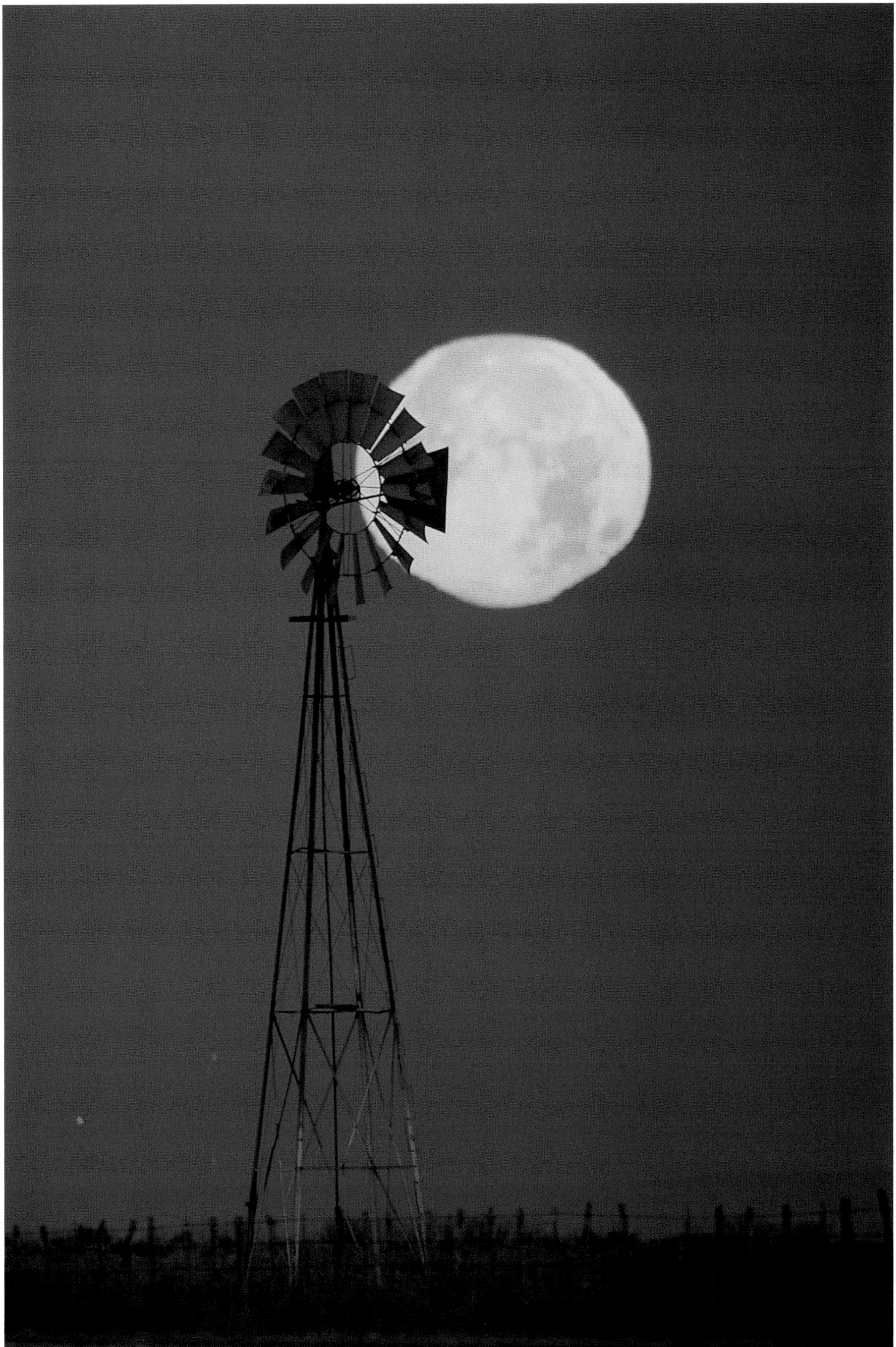

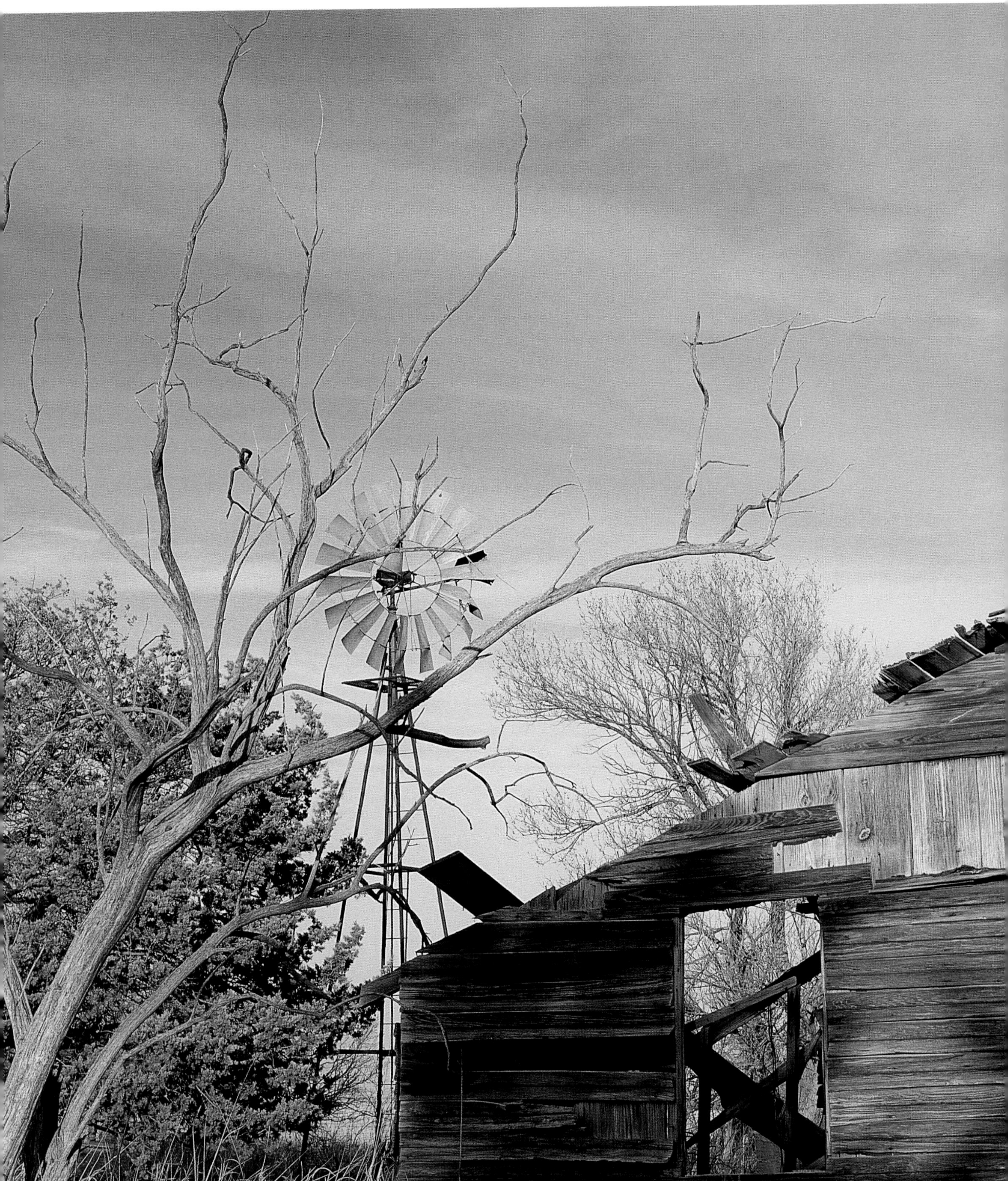

"**M**Y WIFE AND I were new to this country, but we had been told we needed a windmill because it was so dry. I had ordered a house from the Sears catalog before we left Ohio, and it arrived about the same time as the windmill. Some neighbors came over to help put it up, and my wife cooked several meals for all of us. I remember that windmill as being a 602 model Aermotor, one of the best a person could buy then. It took about two months to build the house. The windmill didn't take that long to put up, so we used the water from that mill while we worked. If it hadn't been for that windmill we never would have made it. I took care of that Aermotor, and when my wife died and I moved into town, that windmill was still there, pumping water for the new owners."

ONE LADY REMEMBERS windmills standing by the side of the road between Matador and Paducah, Texas. "There were windmills with tanks in the field for cattle and one near the road for travelers."

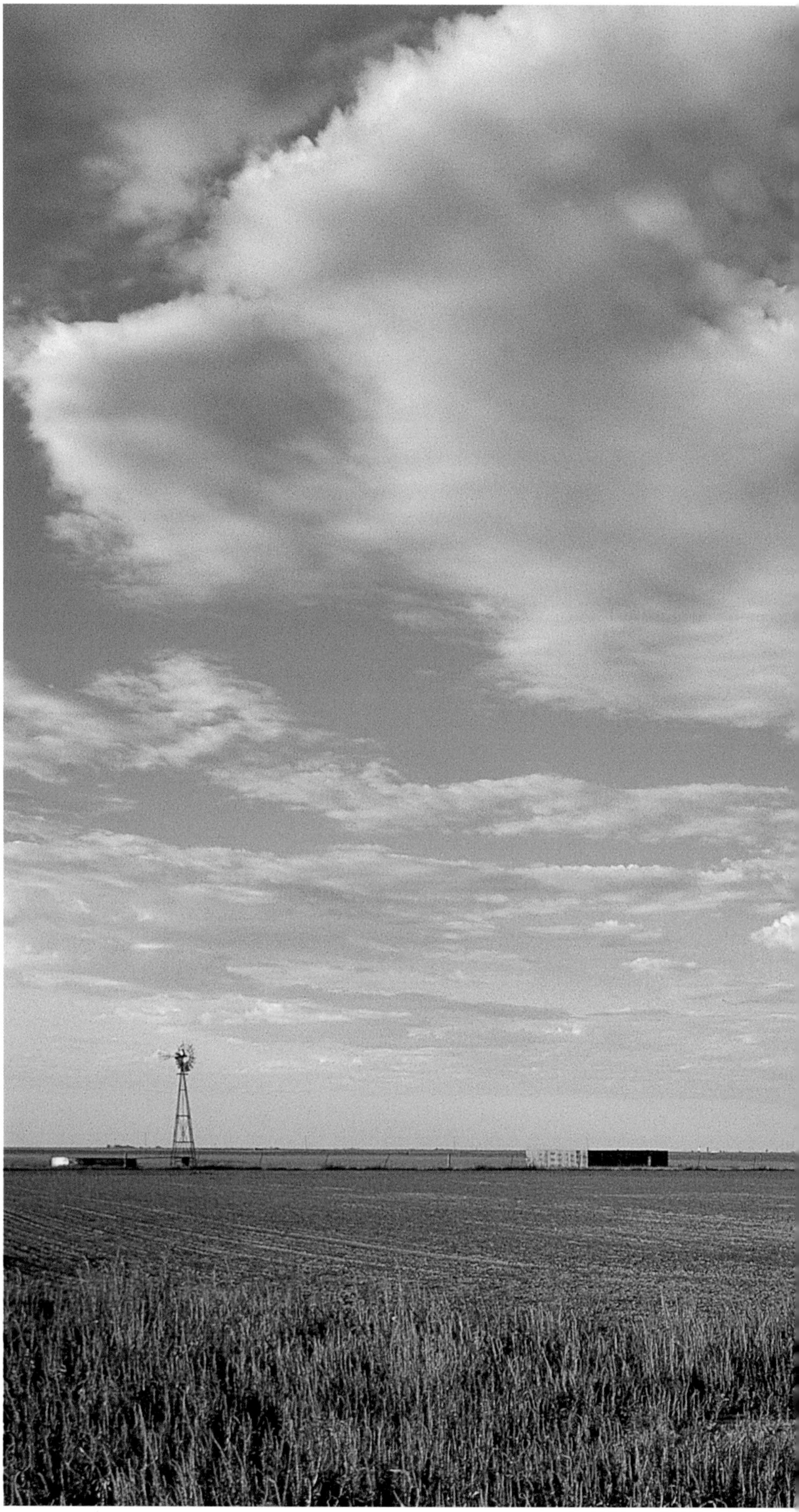

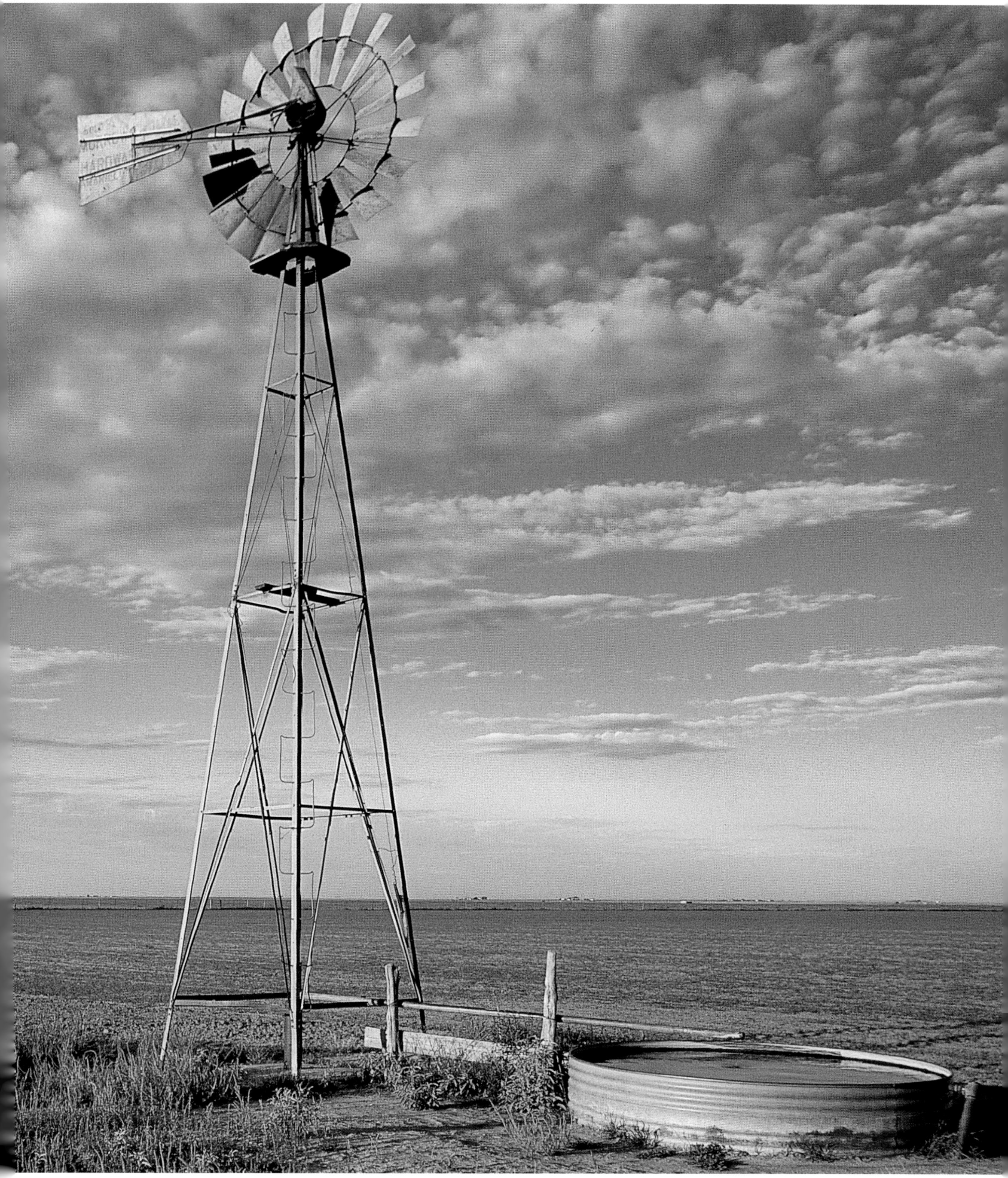

A MAN VISITING THE MUSEUM said that as a child it was his job to churn the butter. He really hated churning, so he figured out that if he tied the churn handle to the windmill's sucker rod, which was going up and down, the windmill would do the work for him. This worked out really well, and they had "windmill butter" whenever the wind was blowing.

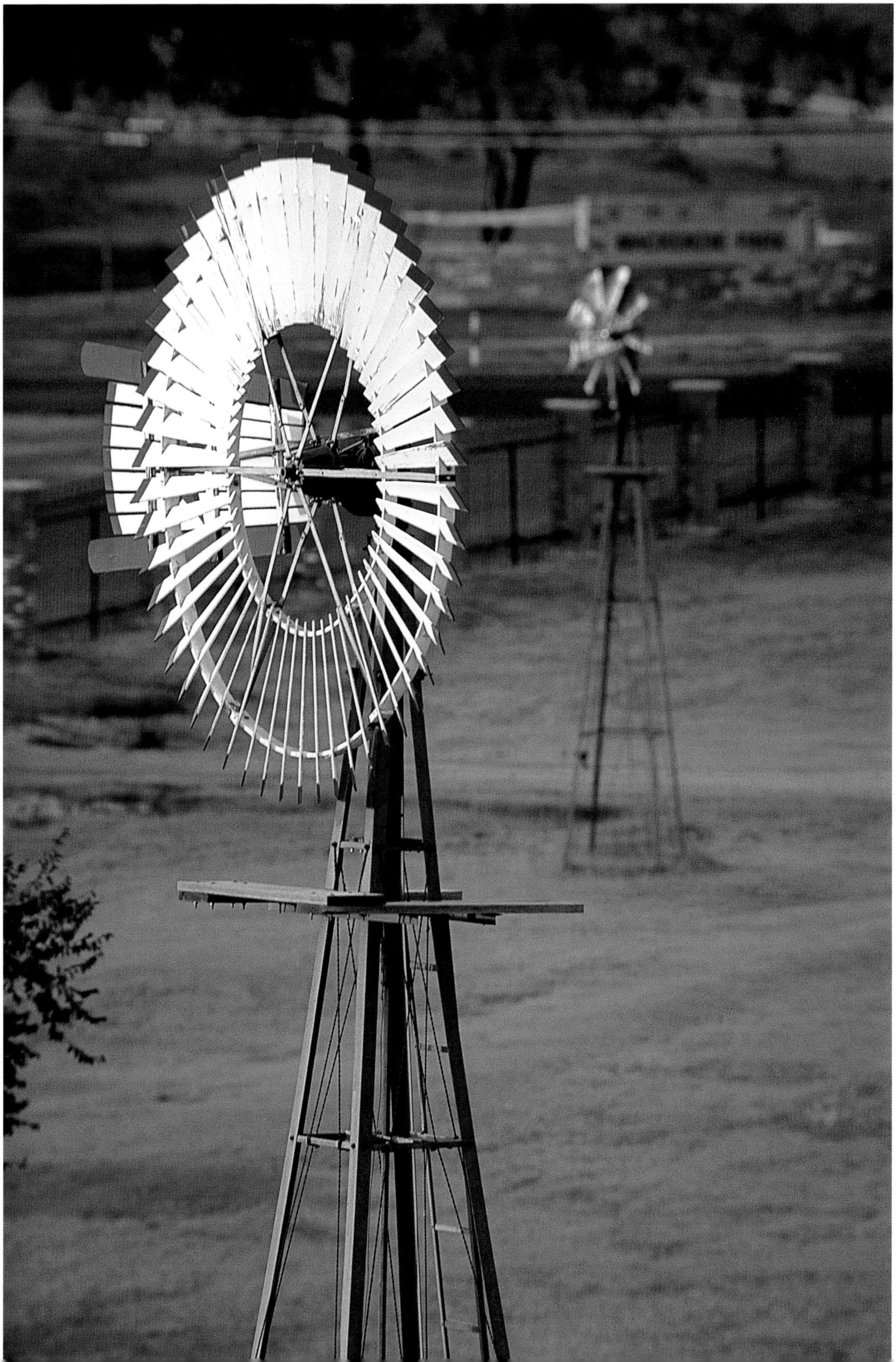

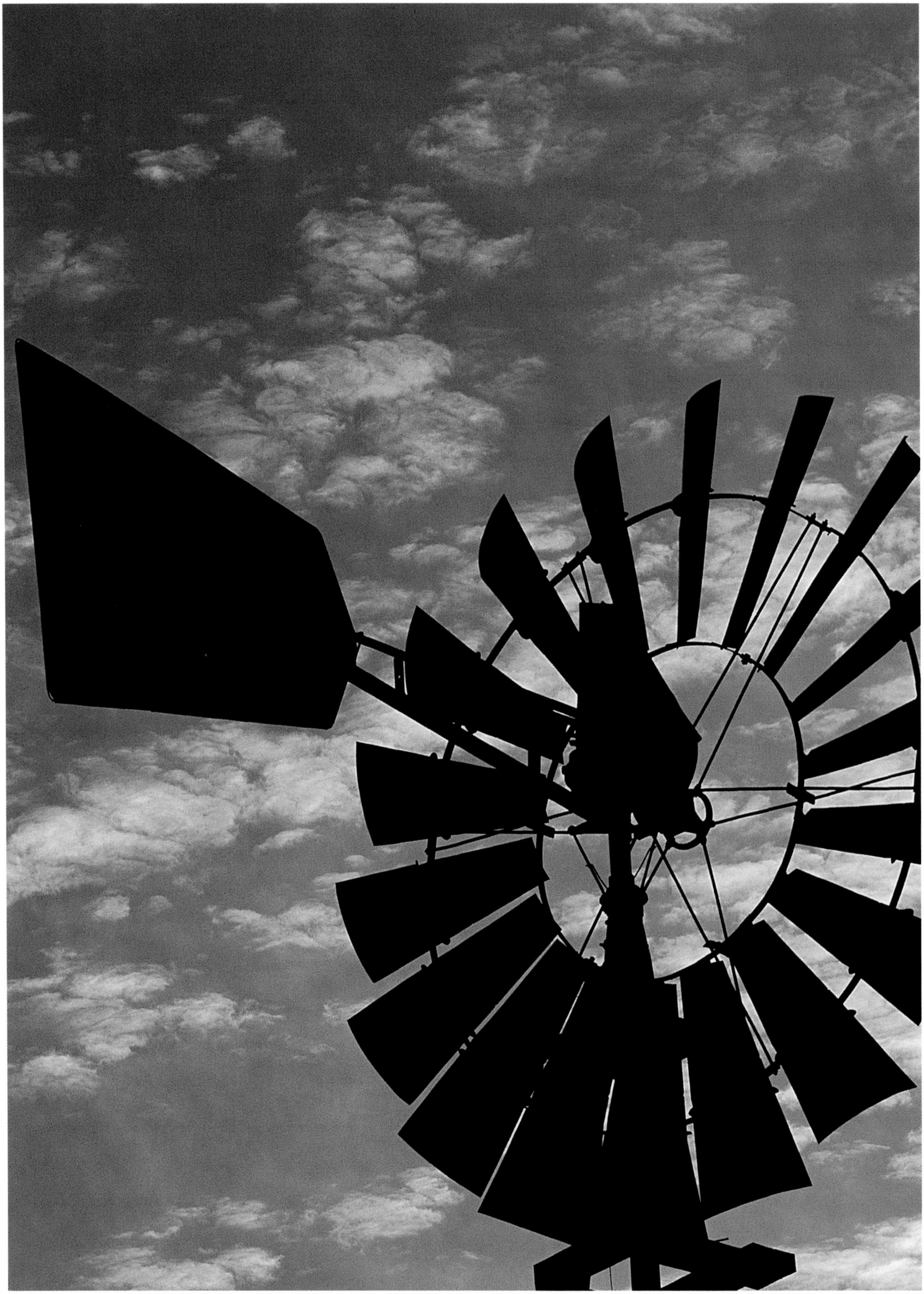

A VISITOR TOLD ABOUT HIS GROWING UP with an Aermotor windmill. He said it was a very large mill, but to him, a small child, everything looked big. One day he and his brother decided they would climb the tower and step out onto the blades of the wheel. They planned to hold on tight and wait for a wind to blow through and turn the wheel. They didn't notice, however, that a squall was moving in. Just as soon as they got "set in," hands holding tight and feet locked into the rim, the wind came up and rain started to fall. He said he doesn't remember just how they managed to get off the spinning wheel, but he does remember thinking he was going to die.

"**W**INDMILL TOWERS WERE USED FOR MANY THINGS besides holding up the windmill. On our farm when a freshly butchered side of beef was available, a pole would be poked up inside the tower, stood up, and tied off without touching the sucker rod. After wrapping the beef in a clean cotton sack, we would hang it on the pole so the beef would cure. This would take a couple of weeks. When you needed some beef you would go cut off the thickened skin for some of the best beef jerky ever, and it would last most of the winter. All you had to do was just 'go cut some meat off the windmill.'"

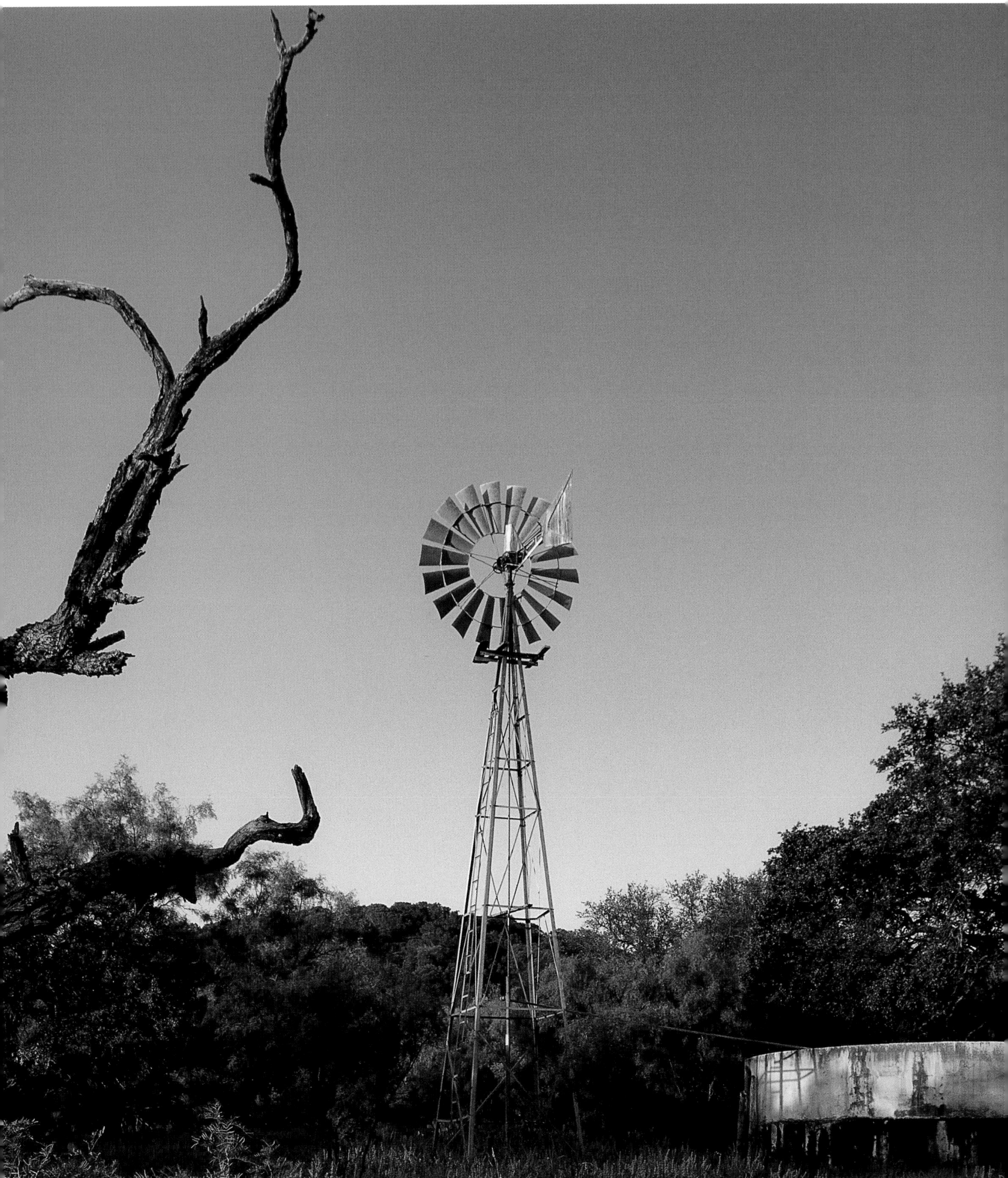

IT WAS THE JOB OF A MAIL CARRIER in Nebraska in the 1940s to fly the mail from his home to Colorado every day. One winter they had a big blizzard that lasted several days. After the storm no one could drive, but it was possible to fly. He took off for Colorado, and on his way he noticed a man out digging in the snow. On the return trip he flew over that same fellow still digging away. He found a place to land and asked the man if he needed any help. The fellow told him he was okay, but now that the snow had drifted higher than the windmill, he was digging down to find the gear case cover to change the oil.

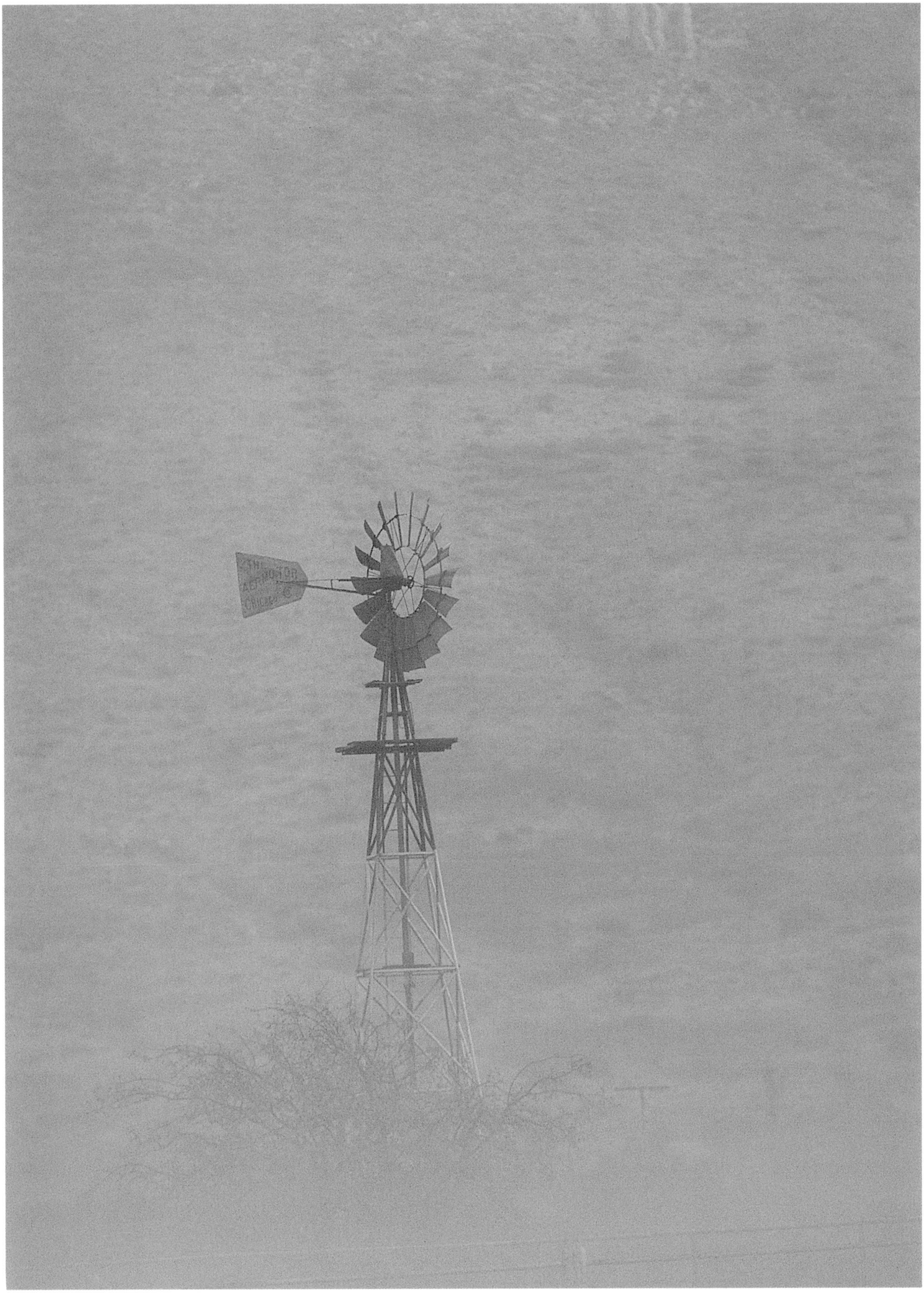

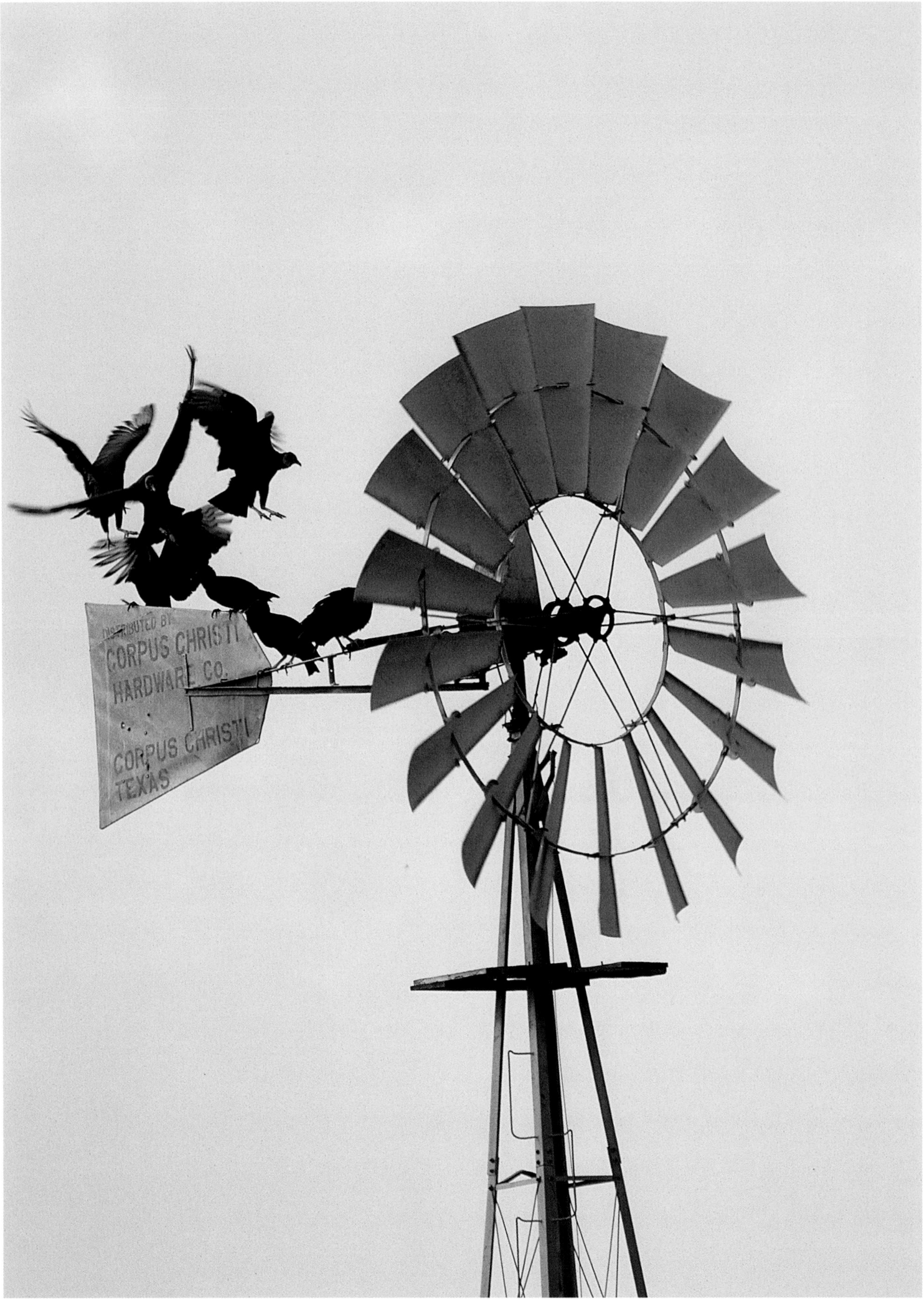
CORPUS CHRISTI
HARDWARE CO.
CORPUS CHRISTI
TEXAS

IF YOU LEAVE YOUR WINDMILL TURNED OFF for a long period of time, the birds, in southeast Colorado, at least, will build nests in the ironwork. One day Rick brought in a nest that he had taken from the top of an old Dempster mill. A raven had worked several summers scrounging up bits and pieces of wire and had woven a big nest entirely of wire. Many people who visited the museum said that a bird couldn't do such a thing, so Rick brought back a second nest still firmly planted in the top of a tower. He set up the tower with the whole nest still in it as a second example just to show people what Colorado birds can do.

"MY MOM AND DAD DID ALL THE WORK on the windmill that was standing by our farm house. It was the only way we got any water, as the nearest stream was several miles down the road. The small lakes that formed after a rain didn't last long, so windmill water was all we had. After a bad storm, my parents had to go fix the mill, and mom was standing at the bottom of the tower where the sucker rod goes into the ground. Dad was working up top and shouted down to mom to 'hold that rod.' When she grabbed for it, the watch she was wearing slipped off her arm and fell down the well. I remember that there was a lot of commotion with mom yelling at Dad while he was climbing down the tower. He tried but couldn't ever fish that watch out from the bottom of the well. We never took that windmill down even when we finally got electricity; mom wouldn't let us. She said that windmill was her 'watch tower.' But Dad did buy her another watch."

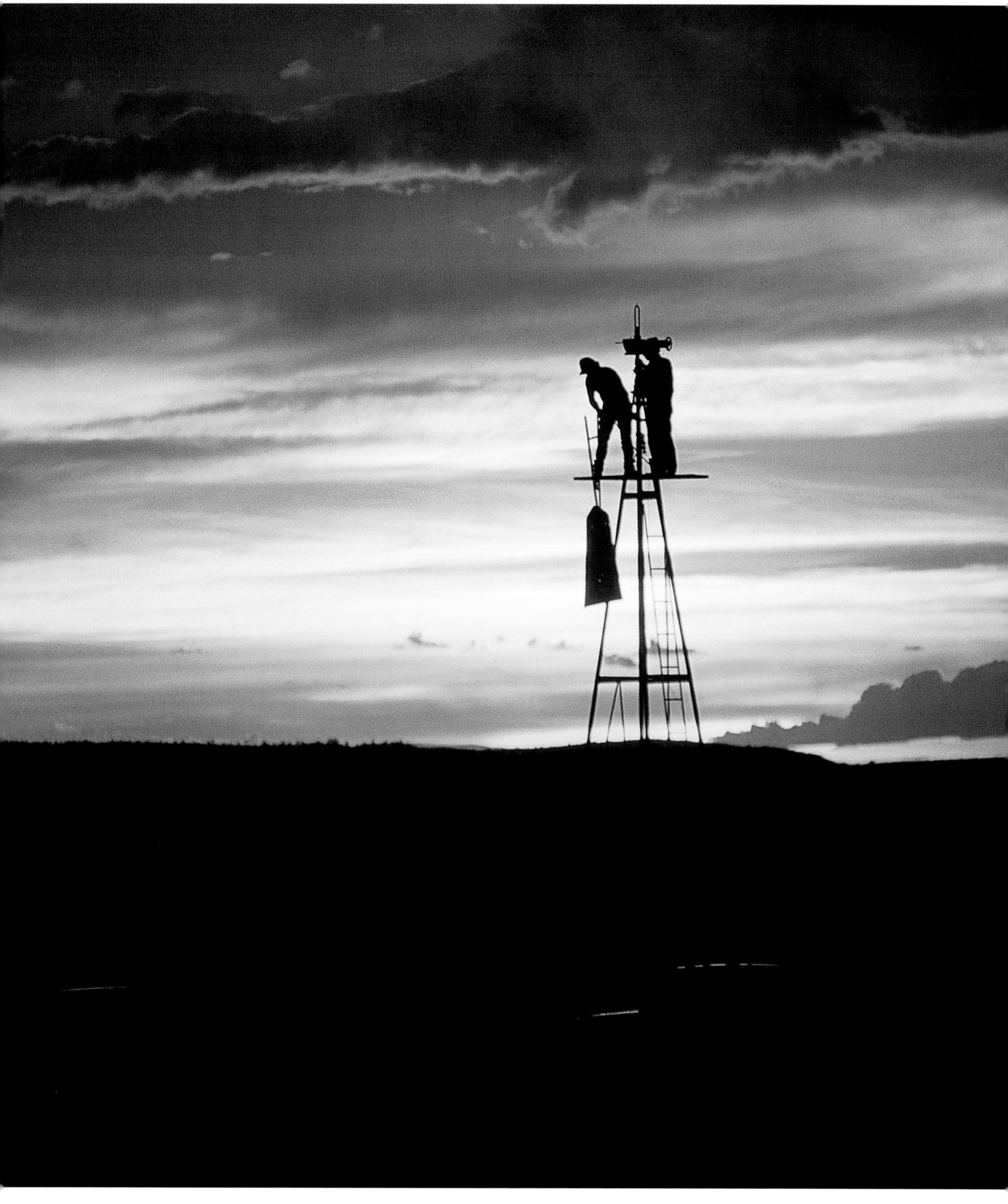

STAN, WHO WAS ABOUT TWELVE, noticed that doves always landed around the edge of the windmill tank. He decided that the best place for shooting doves was up on the windmill tower's platform. He got his gun and climbed the tower, but just as he got to the top and put his head through the hole in the platform, he was nose to nose with a raccoon. He was so surprised that he lost his balance and dropped his gun while struggling to hold on. When he looked again, the raccoon was still there but hadn't moved. Stan poked it and realized it was almost dead. Apparently the raccoon had climbed to the platform but couldn't figure out how to get back down and almost starved to death. Stan took it home and fed it. It hung around their yard for a while, then finally took off for the wild.

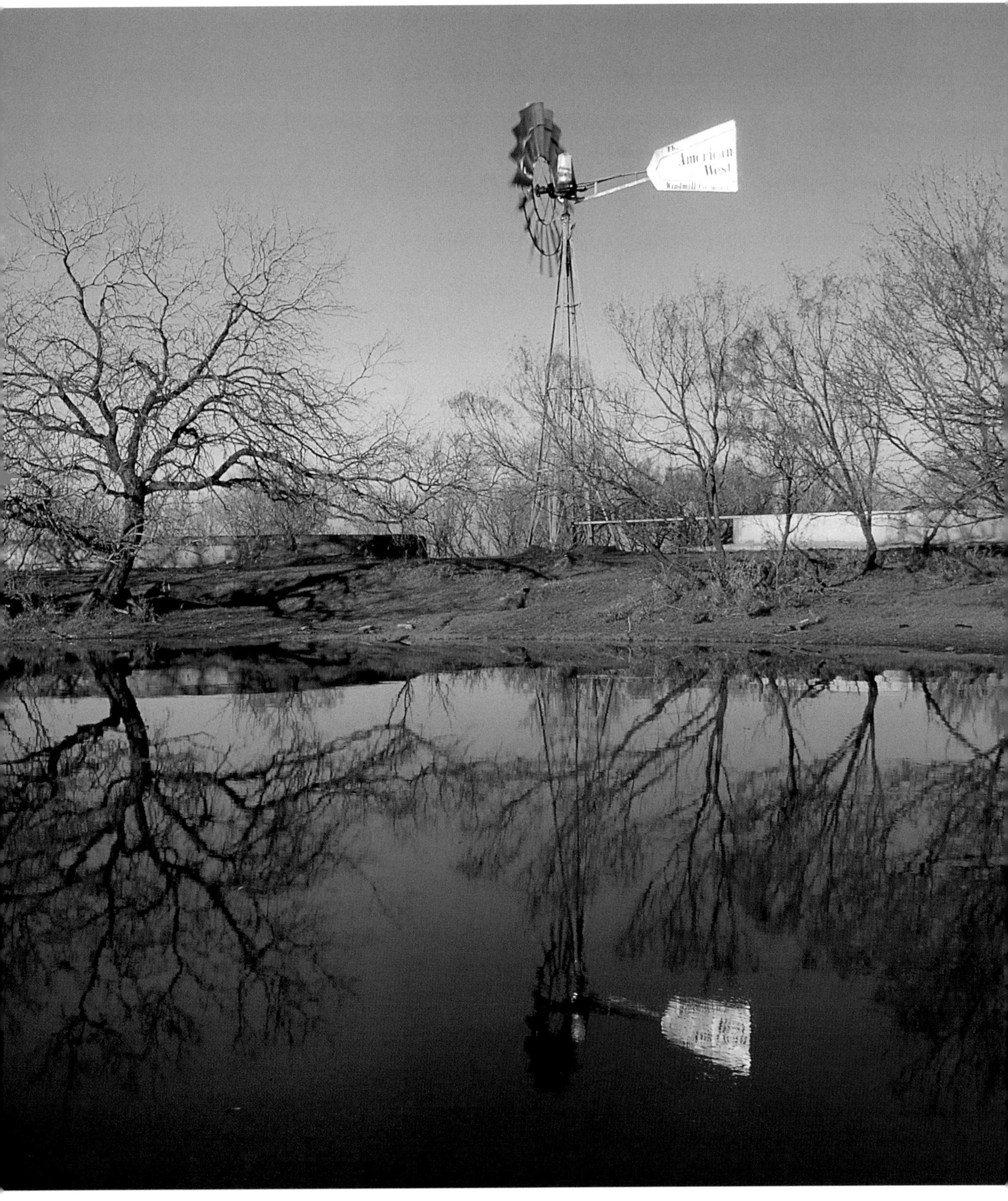
The
American
West

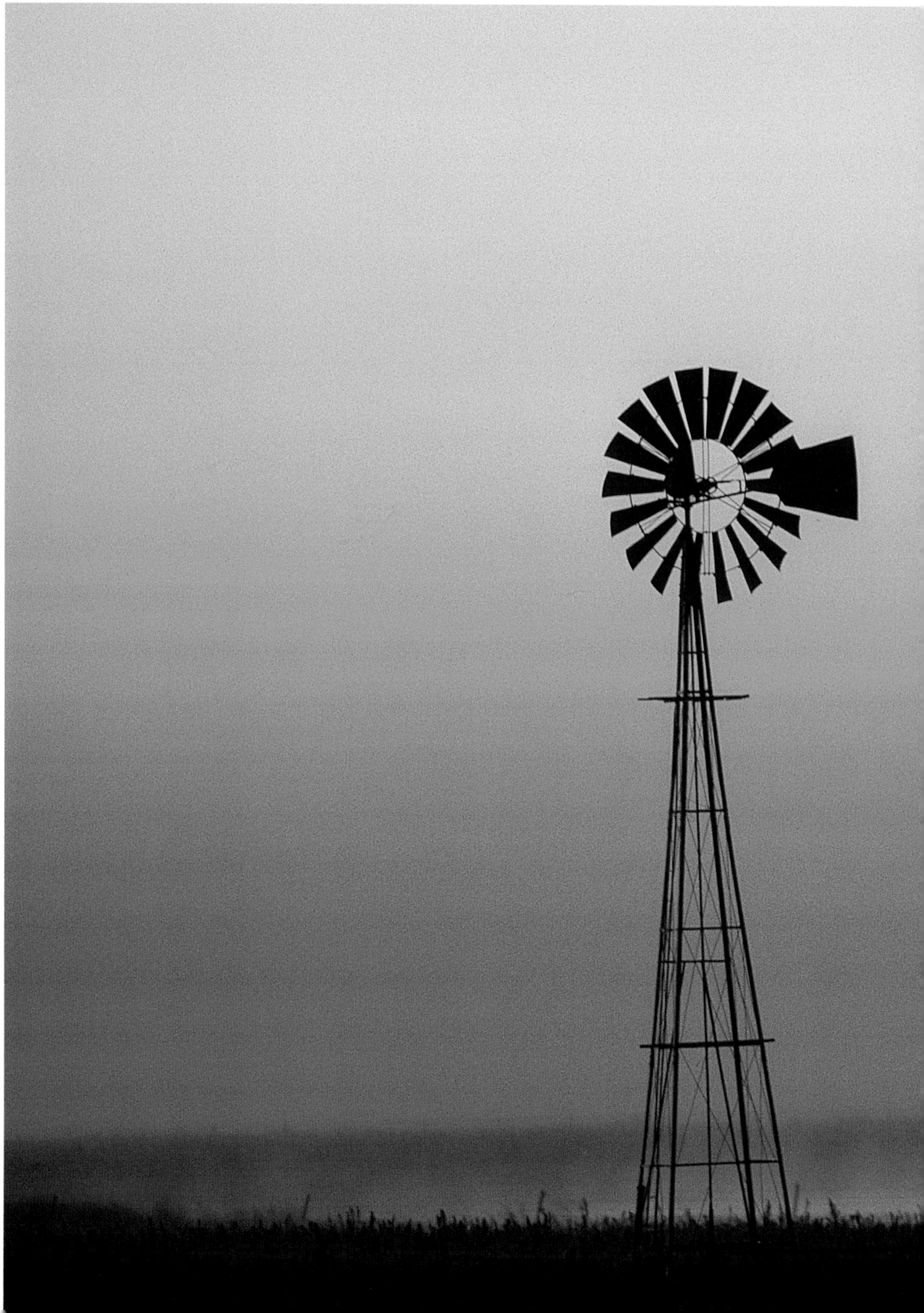

A BEDFORD, TEXAS, WOMAN told how when she was a kid she would climb the windmill tower and slide down the sucker rod. She said if you wrapped a waxed-paper bread wrapper around the sucker rod and held on tight, you could go down really fast.

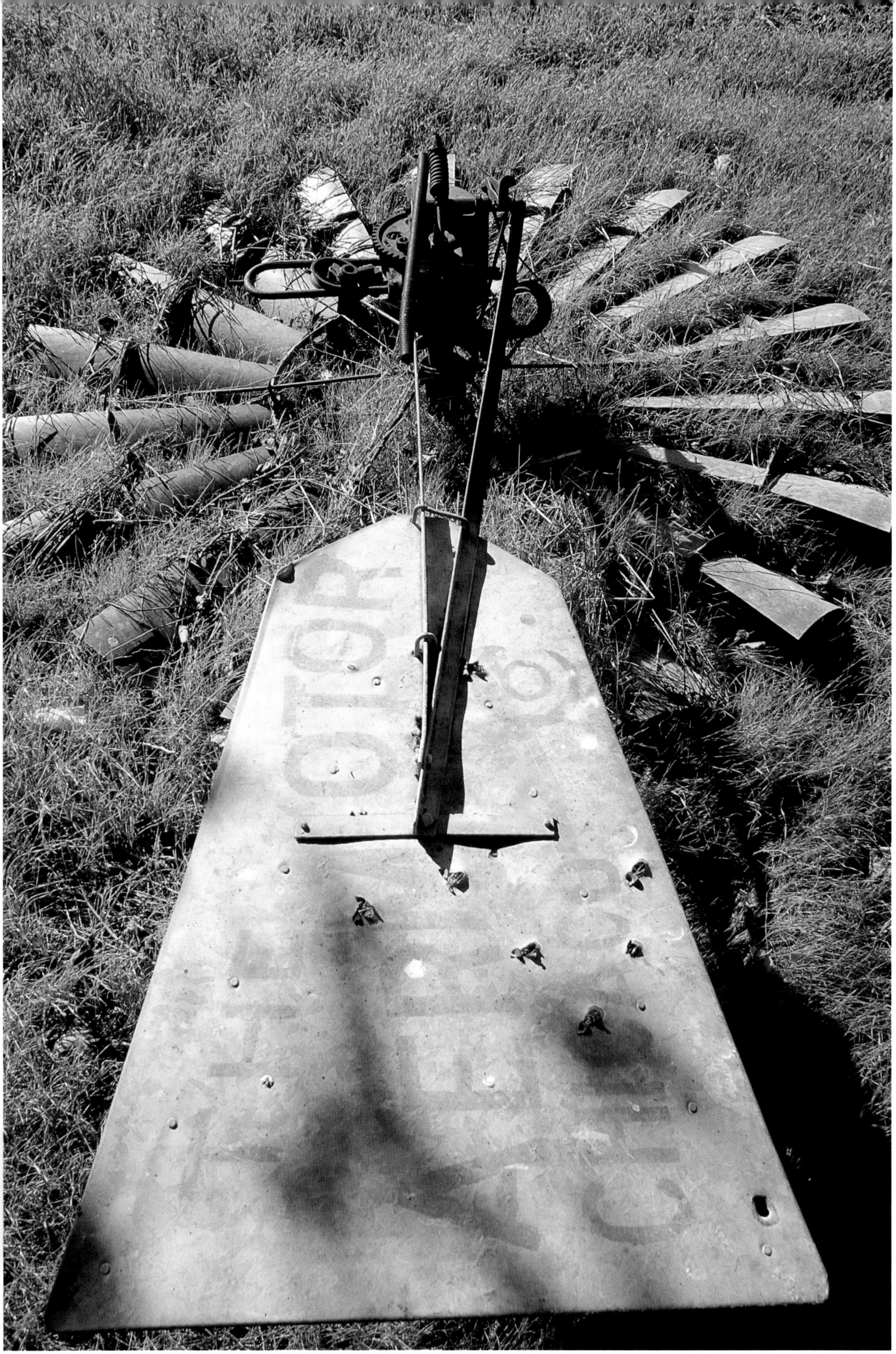

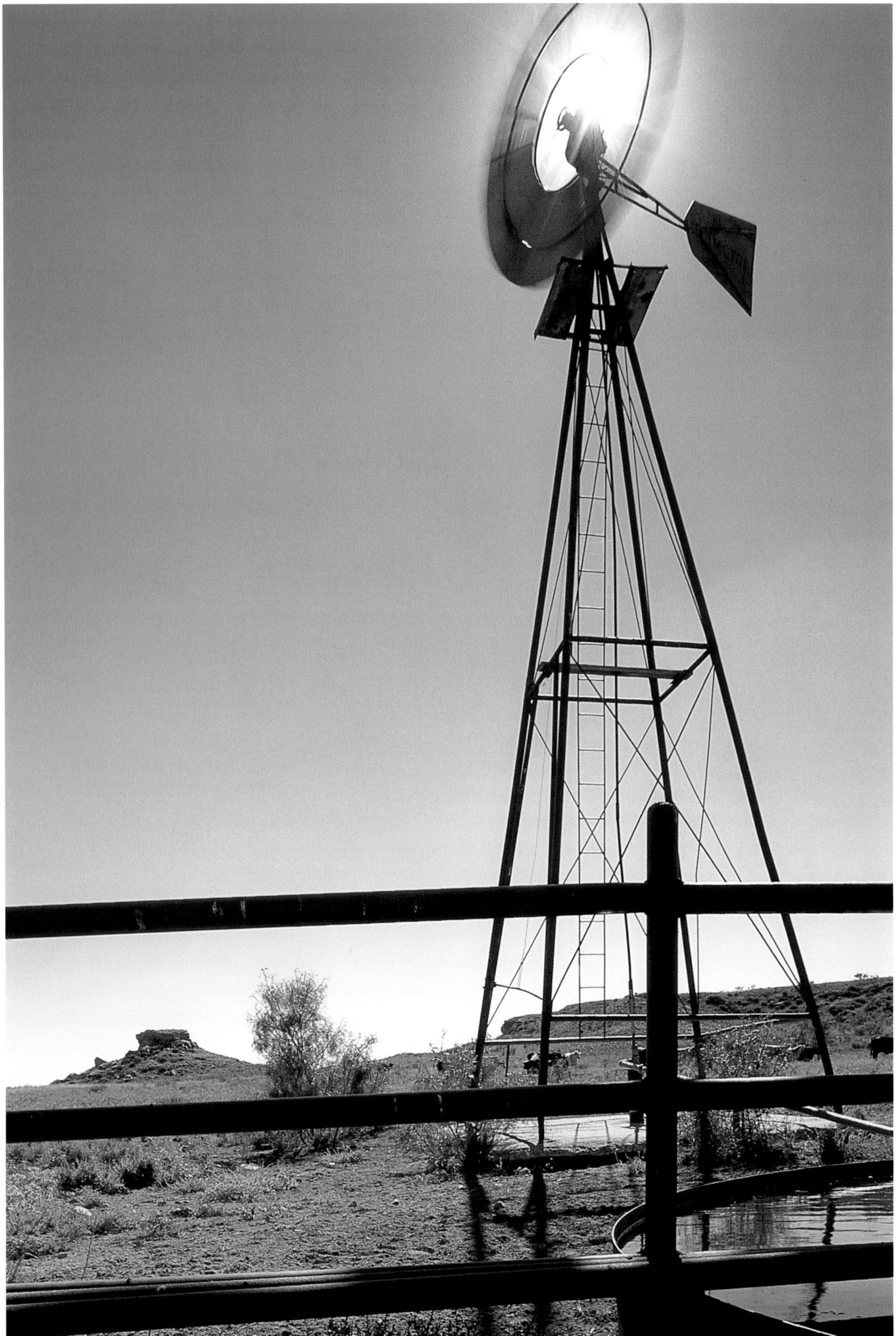

"THERE WAS NOTHING WORSE THAN WASPS, and you always knew there would be nests somewhere near the windmill. One day I was climbing the windmill tower and got up to the platform before I hit a nest with my hand. The wasps just poured out and buzzed all around me. I jumped up hollering, waving my arms, and the next thing I knew I was standing on the ground. I must have jumped right off the platform, because I don't remember anything except those dang wasps."

A FARMER GOT INTO HIS WAGON and started riding off to Littlefield, Texas, late in the afternoon. He was a little apprehensive about having to return down that dark dirt road at night, so he decided that on his way to town he would count the windmills between his house and town so he would know how far to return. After he finished his business and started back home, it got dark sooner than he expected. He realized that he couldn't see the windmills to count them and began to admit that he was lost. There was no moon, and he could barely make out the shape of an occasional house along the way. Finally, he decided to stop at the next house to find out where he was. He drove the wagon down the road to the next farm house and went up and knocked on the door. When the door opened, he was surprised to see his wife standing there in his own house.

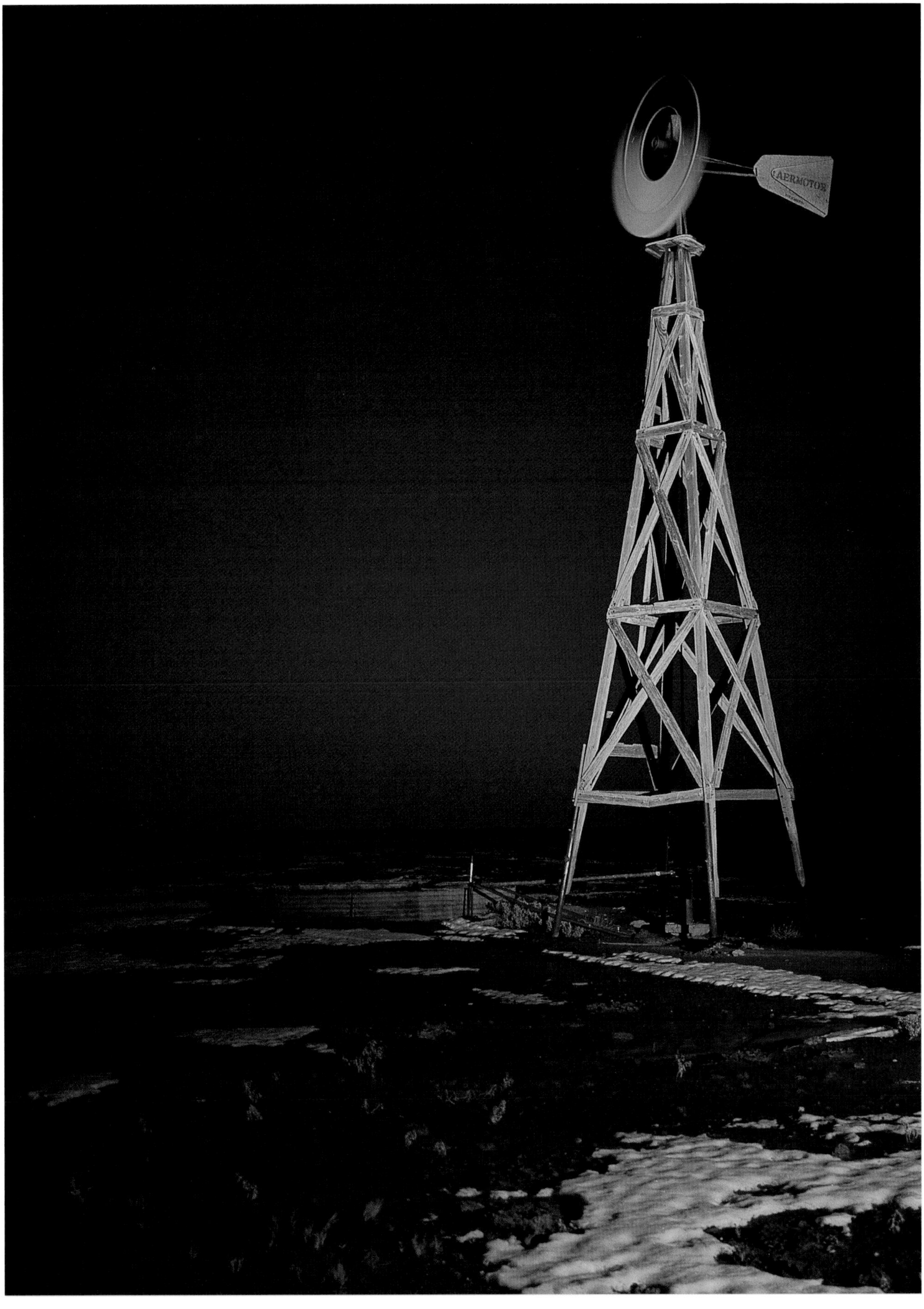
AERMOTOR

ROBERT'S DAD SENT HIM ON HORSEBACK, with sacks of corn tied to the saddle, to have it ground into meal. After a long ride he got to Lockney, Texas, where there was a grain grinder powered by a windmill. Not a bit of wind was blowing, and Robert spent the night and the next day waiting for the wind. The third day came and went, and finally, late that night, they heard the wind start up. Robert jumped from bed and ran to the windmill to grind the corn while the wind was blowing hard. He got his corn ground early that morning and then headed for home, four days after he left.

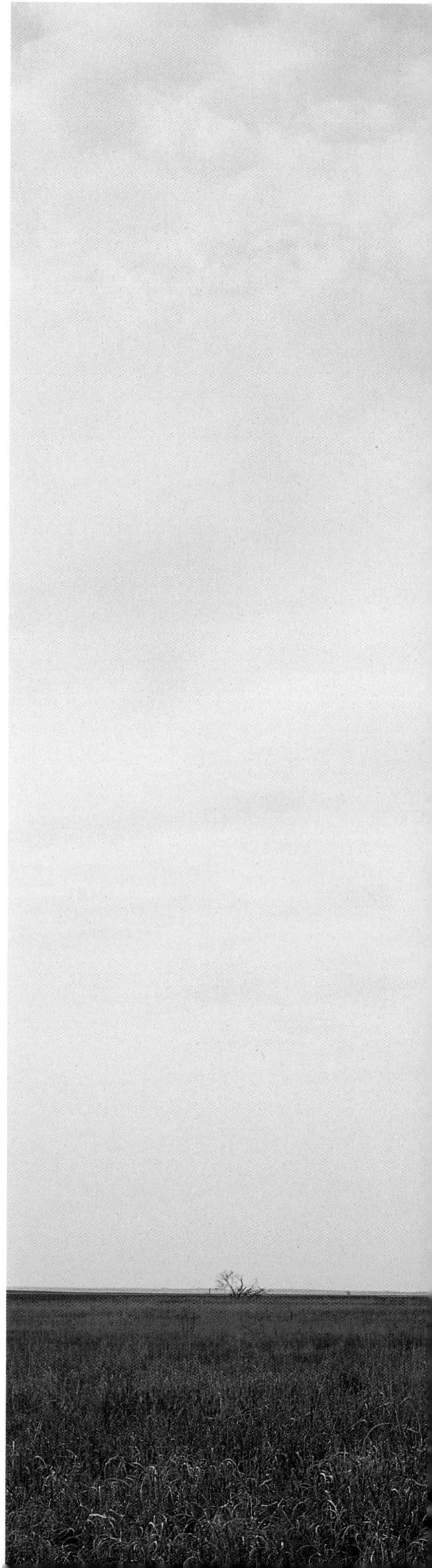

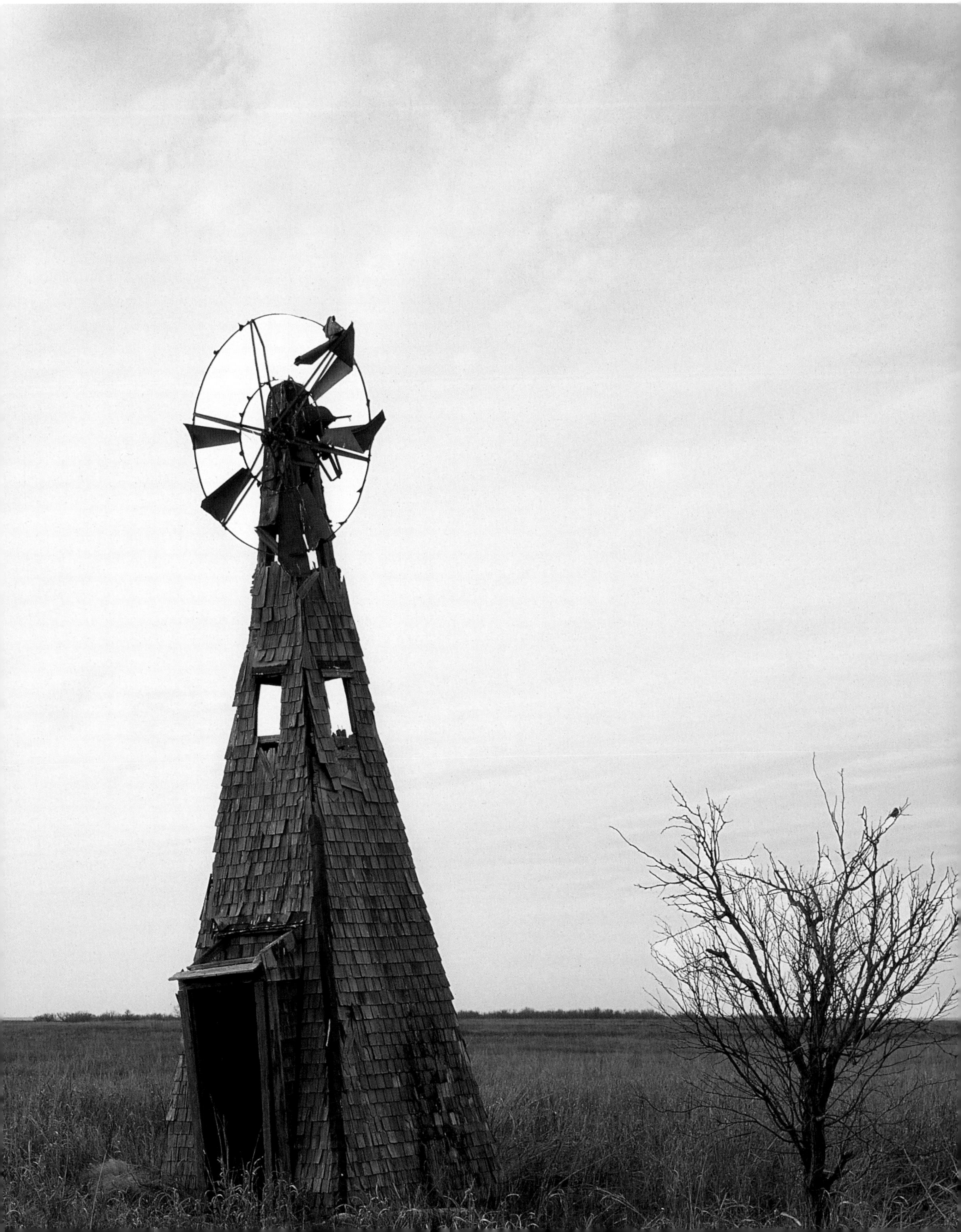

"SUMMER ON THE HIGH PLAINS OF TEXAS seemed
delightfully cool after the humid heat of central Texas. At
the train station in Canyon, my grandfather met us in a
covered wagon and the trip to their ranch may have
required two days. My grandfather was tall and spare,
with a red beard and tight little red curls across the back
of his head. The top of his head was bald. My grandmother was plump and round, with blue eyes
behind round steel-rimmed glasses.

"Near the house was a windmill with a pond
where my grandparents raised catfish. They came like
chickens for breadcrumbs or table scraps, and leaped at
the edge of the water. My grandmother dipped them up
with a net, and we had catfish when we wished. Without
refrigeration, leftovers went back to the chickens, the pigs,
or the fish. My grandmother said the food would
come back to the table, and it did.

"Dust storms were unknown then; it was a world
of fresh breezes, wide windy sky with scudding clouds,
green grass, and shallow rippling lakes,
windmills and cattle."

*—from a remembrance of Tyline N. Perry,
talking about trips to her grandparents' in 1900.*

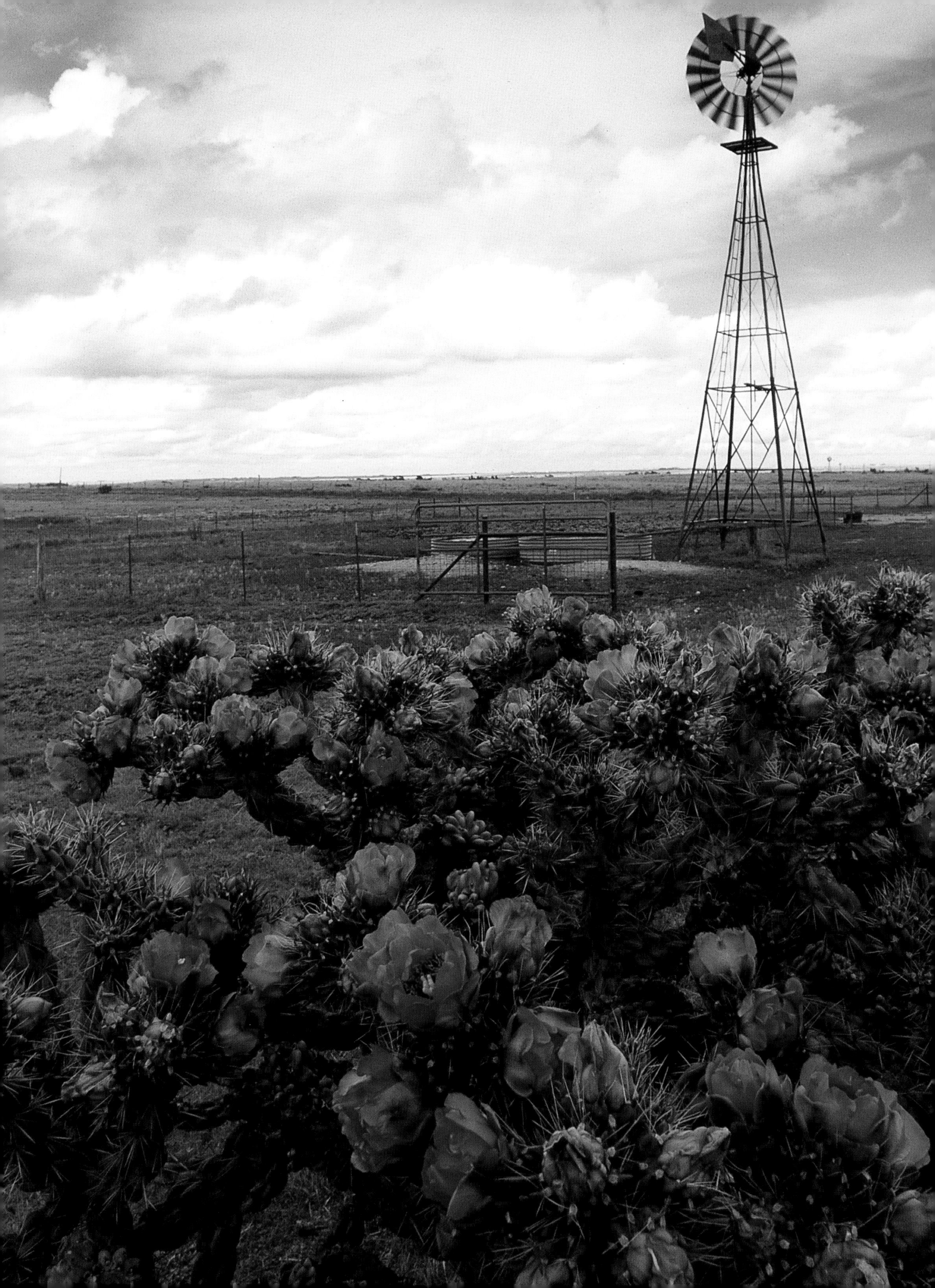

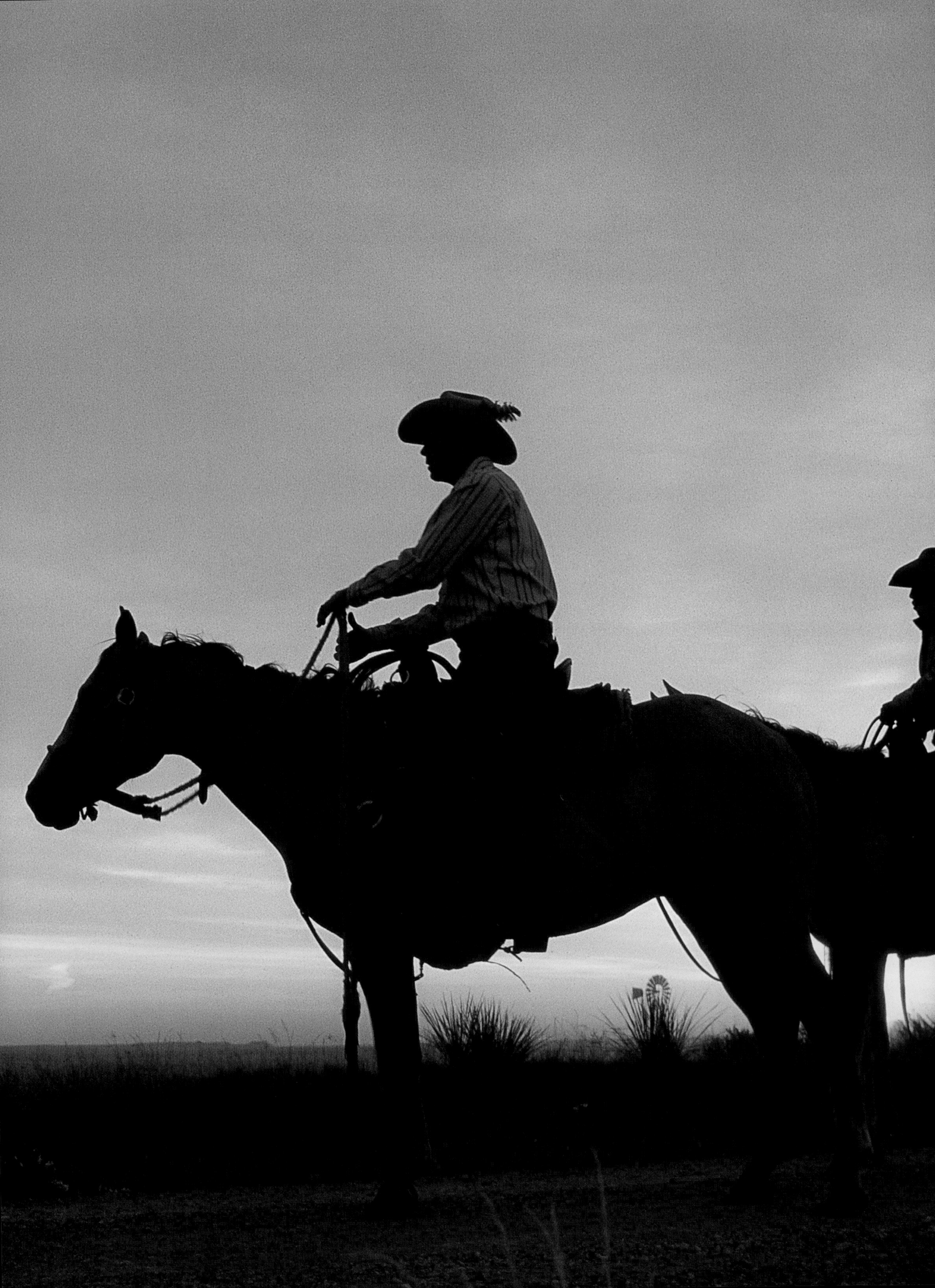

In northern Wyoming some cowboys worked ranches that were so large a nearest neighbor might be seventy miles away. It was always a real treat to be invited to someone's house for a meal, but especially so to a certain rancher's house. That ranch had a Wincharger, mounted on a skinny tower, which generated electricity from the wind. The cowboys said they had their best meals when they got to go there on Saturdays at noon. After the meal they listened to the "Texaco Opera" on a radio powered from batteries charged by the Wincharger.

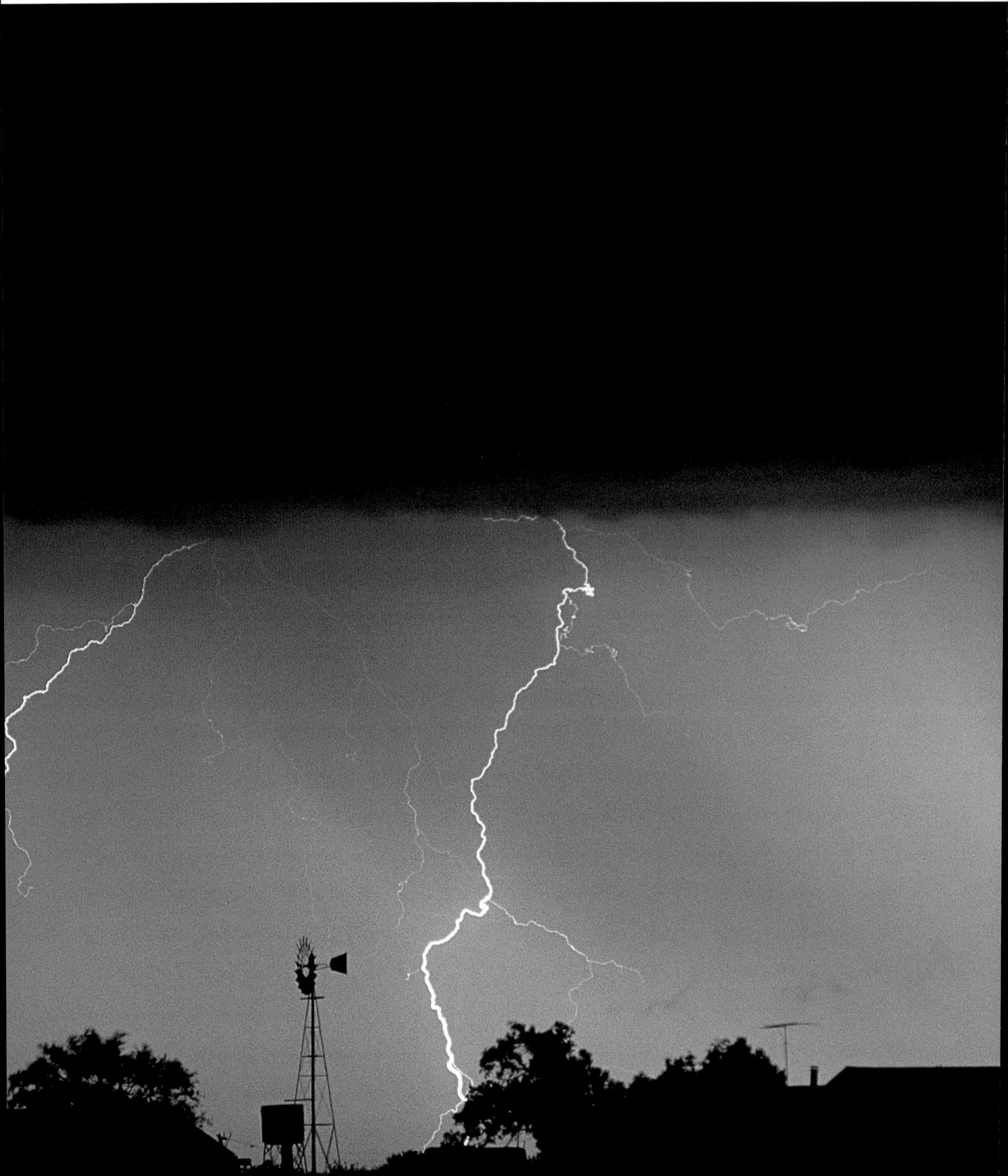

"WE WERE STANDING AT THE BOTTOM OF A BRAND NEW TOWER we had just put up, and on top was a bright, shiny Dempster windmill. Cows had started to come up to get a drink of water and were standing around the empty tank. We had been watching the clouds while we worked and heard some thunder off in the distance. I didn't think too much about it until I looked at my partner and saw his hair standing straight up on his head. He was looking at me. I didn't have much hair, but he later said it was sticking up around my ears. I knew what was going to happen next and yelled, 'Let's get in the truck now!' Right then a bolt of lightning struck and knocked us both to the ground. Fortunately it didn't hit us and didn't even hit the windmill, but it did strike one of the cows standing by the tank, and it killed her dead. We were both scared and shaking, and it took some time for my buddy's hair to fall back down."

IN THE PASTURE BEHIND THE HOUSE the Whitmore family had a windmill that was their only source of water. Mr. Whitmore said all the time he was growing up, someone was telling him, "Earl go get some water," or "Earl go turn the mill on so it can fill the tank," or "Earl go turn the windmill off 'cause the tank is running over" or one of several other windmill-related chores Earl had to do. He said his whole childhood was spent running back and forth between the house and the windmill.

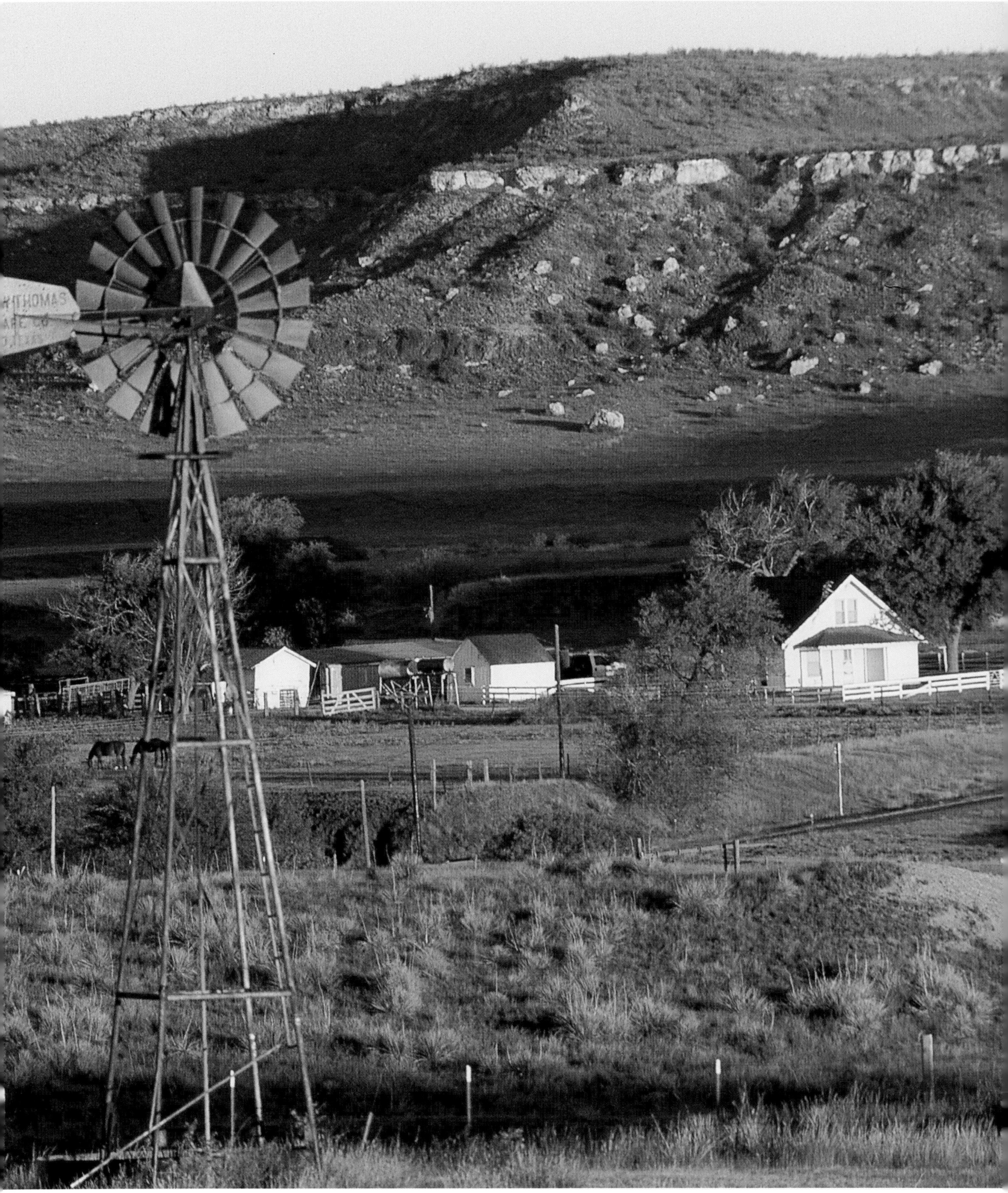
THOMAS

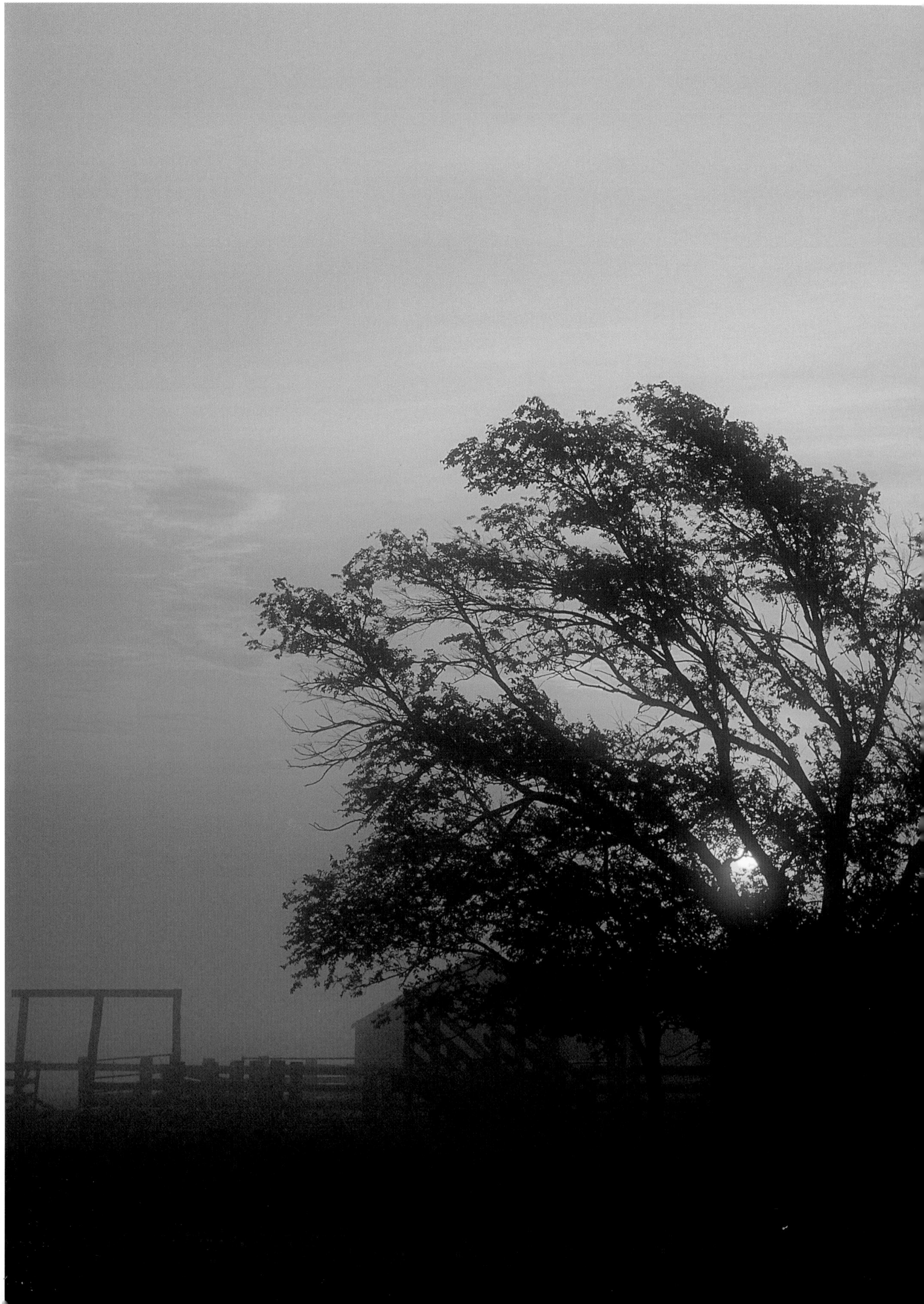

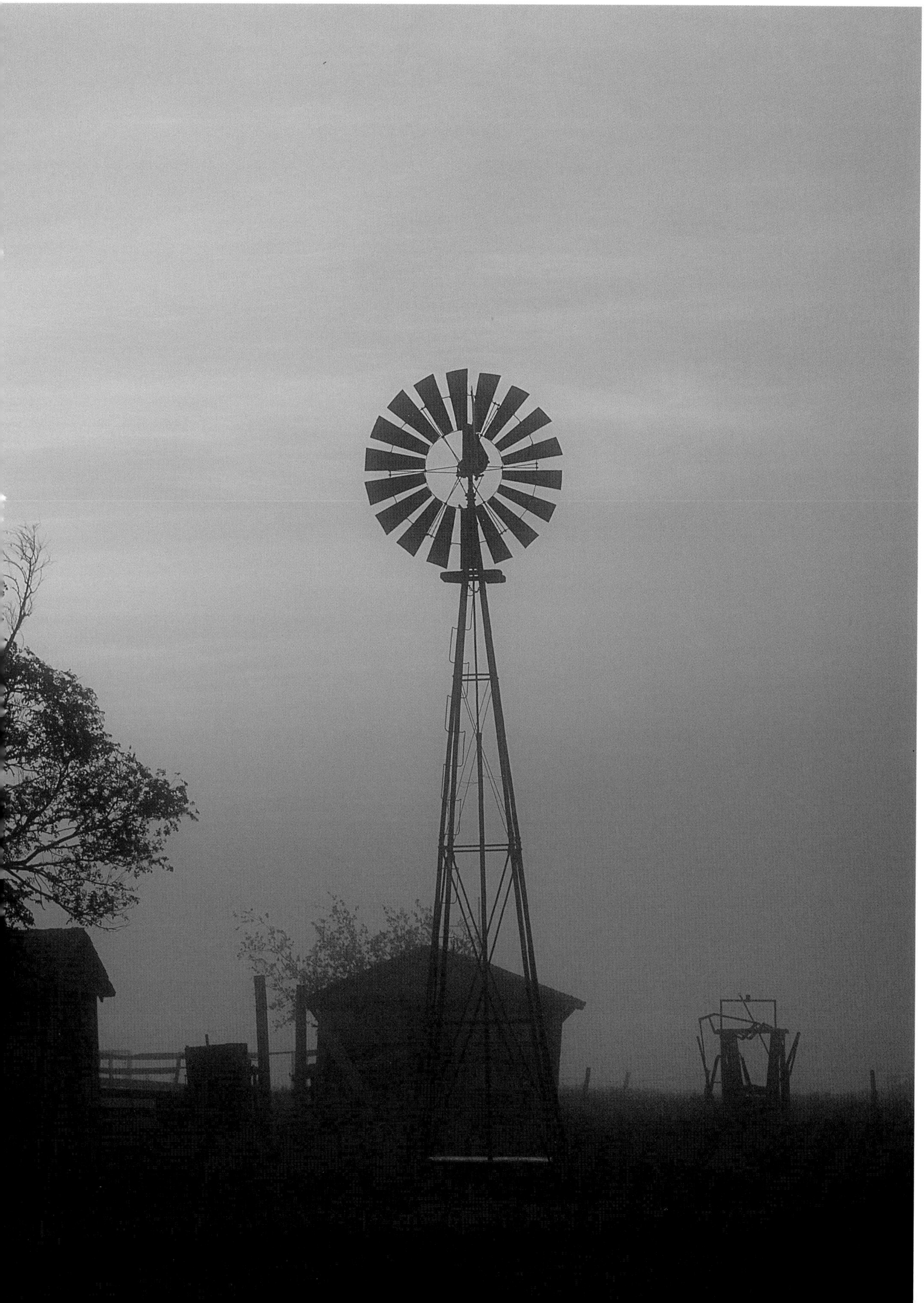

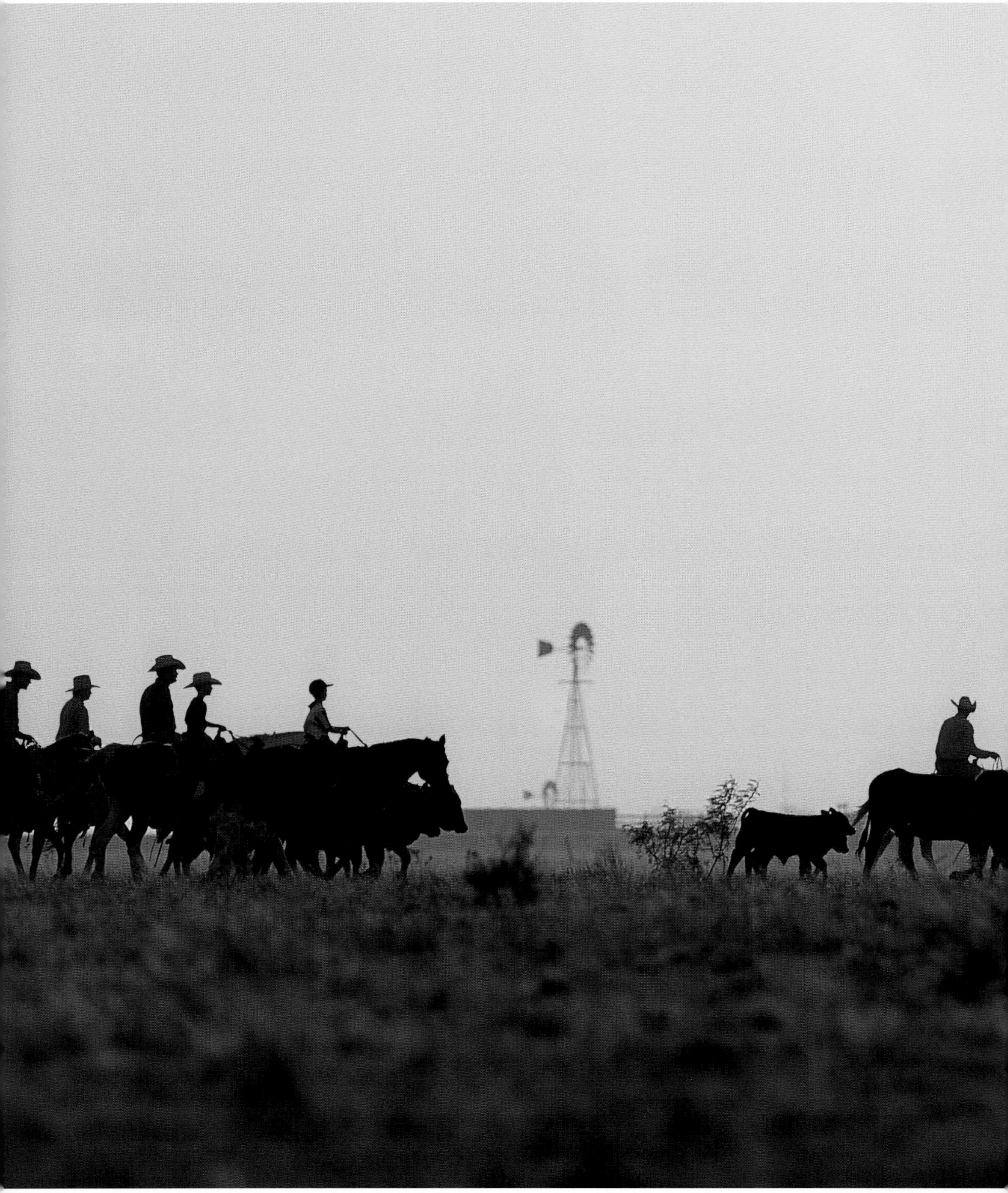

"**F**OR THREE DAYS the wind hadn't blown, and all of the stock tanks were about empty. The cows were thirsty, so my dad told me and my two brothers to go out to the windmill and pump the well by turning the windmill wheel. He said for us to keep pumping until the tank was full. We three boys worked taking turns at the wheel, turning it enough so that there was about one inch of water in the tank. But every time we got it to about that level, the old cows would come and slurp up that water, making so much noise we could hear it on top of the tower. We worked all night, and by morning there was still only one inch of water in the tank. We kept working until we felt a fresh breeze come up with the sun. Then the windmill started pumping by itself, and we were delighted to get off that tower."

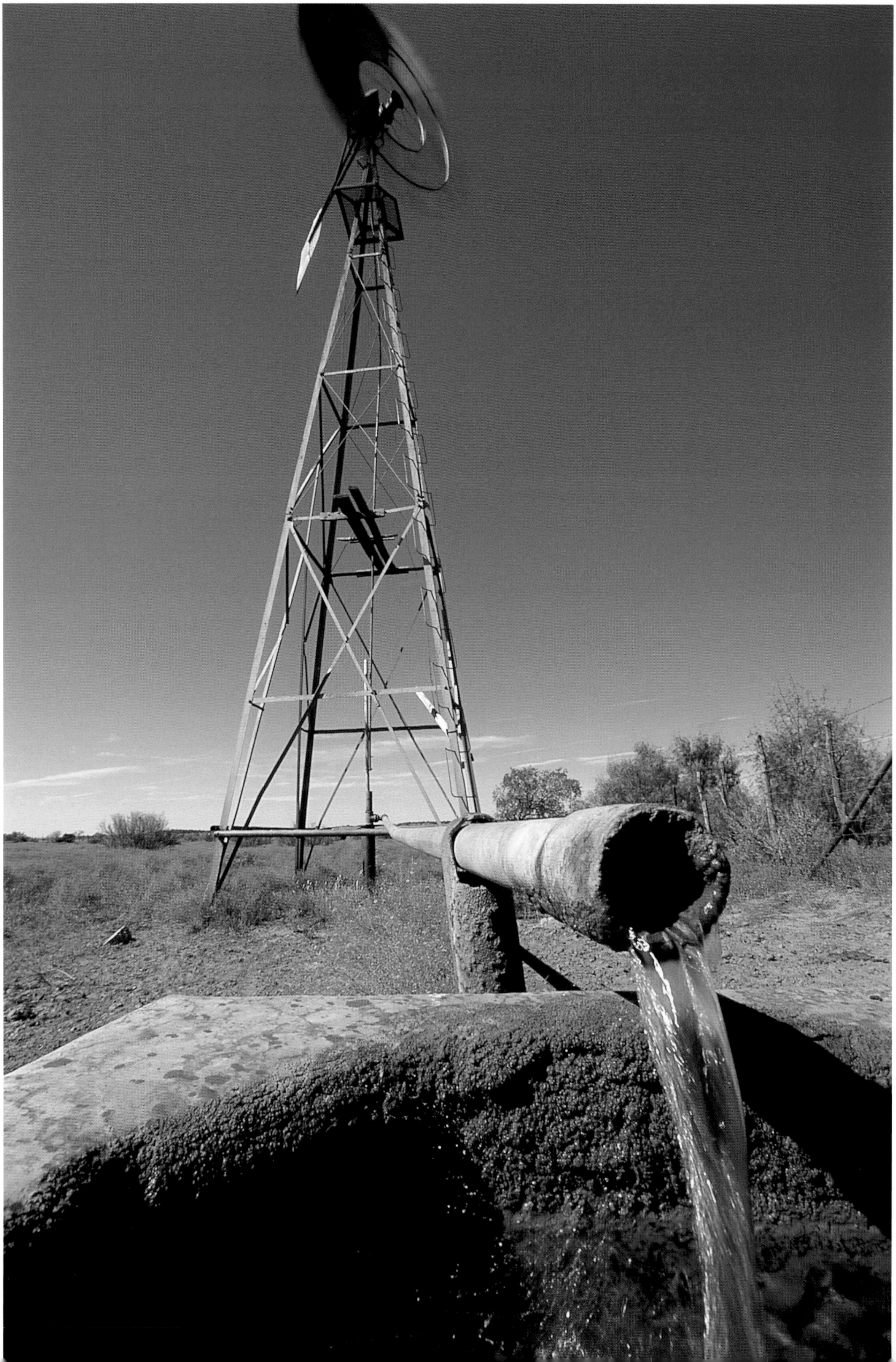

"OUR FAMILY DIDN'T HAVE MUCH MONEY, so we had to do all the windmill repairs ourselves. When my dad and I would change out the leathers on the windmill, if we saw a spot where the drop pipe had rusted through we would wrap a strip of an old inner tube around the pipe and secure it with baling wire. When we dropped the pipe back in the hole, it always seemed to work."

WHEN THE WINDMILL MUSEUM WAS BEING BUILT, the local Texas Department of Corrections facility sent out work crews to help build the fences, walkways, and other structures. Those work crews always wore white uniforms, and one or two guards watched over them. The museum is located on park grounds, and the surrounding area is popular with joggers and bicyclists. One morning one of the museum staff members spotted someone dressed in white running like mad down the road outside the fence. He called out to the nearest guard that one of the prisoners was escaping. Two guards jumped into their truck and caught up with the escapee. They came back somewhat red-faced to report that it was just a local jogger who happened to be wearing white clothes that morning.

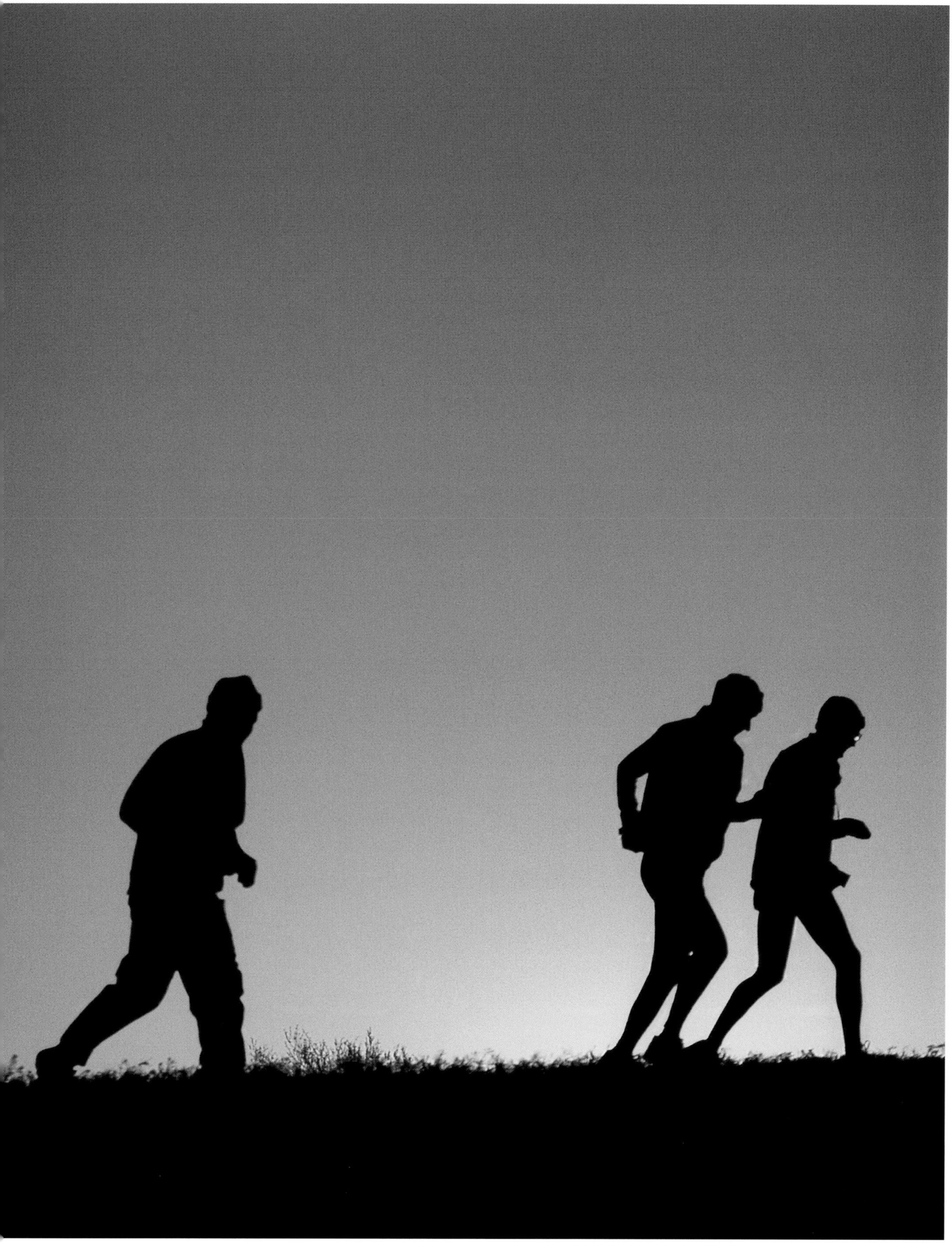

GLENDON STOKES had a younger, six-year-old, red-headed brother who was always getting into things. One day when the wind had stopped blowing, to keep his little brother out of trouble his mother gave him a bucket and told him to go to the windmill and pump water until the bucket was full. As he struggled to get the bucket full, he kept moaning, "I am just wasting my life away."

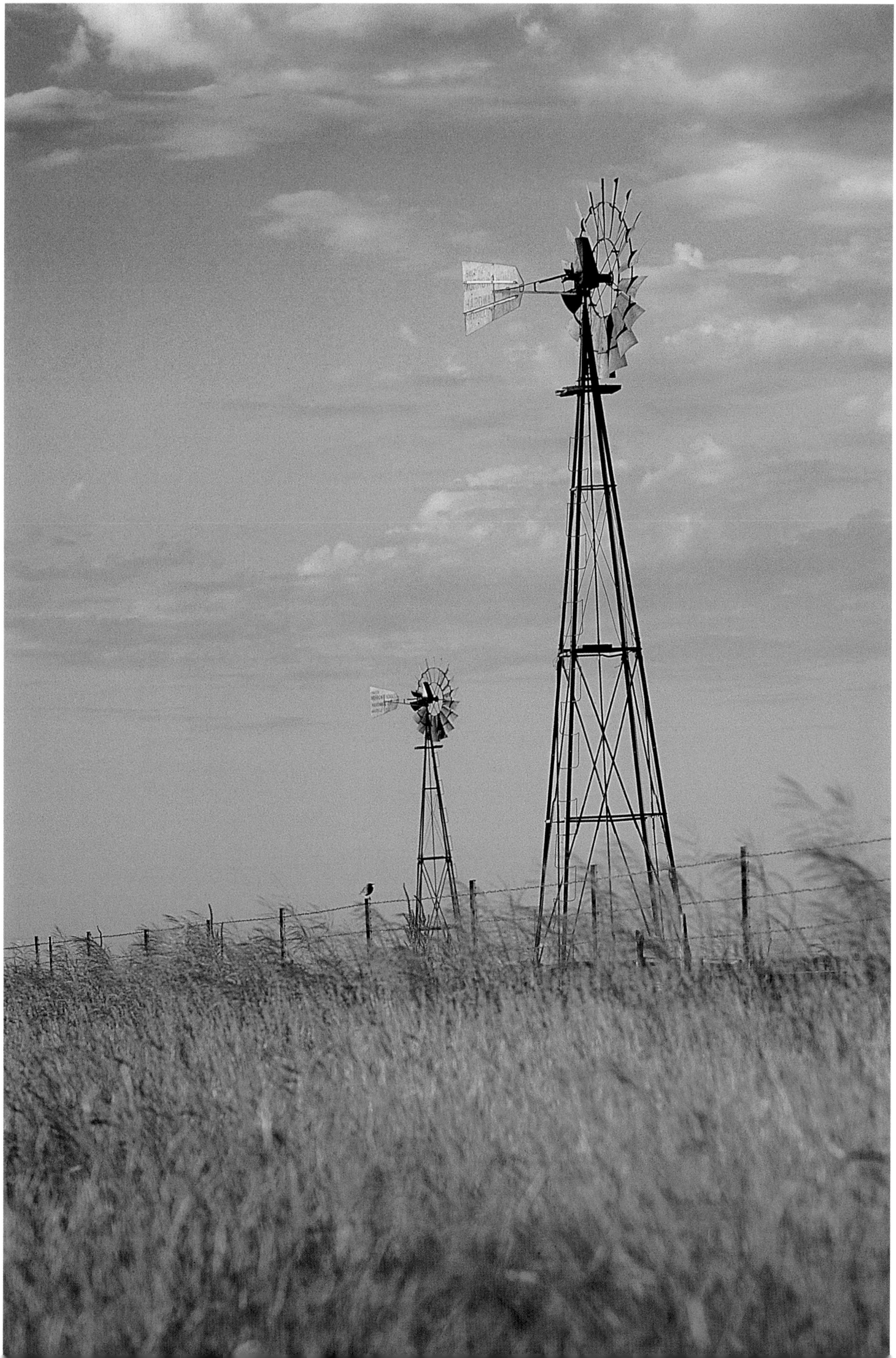

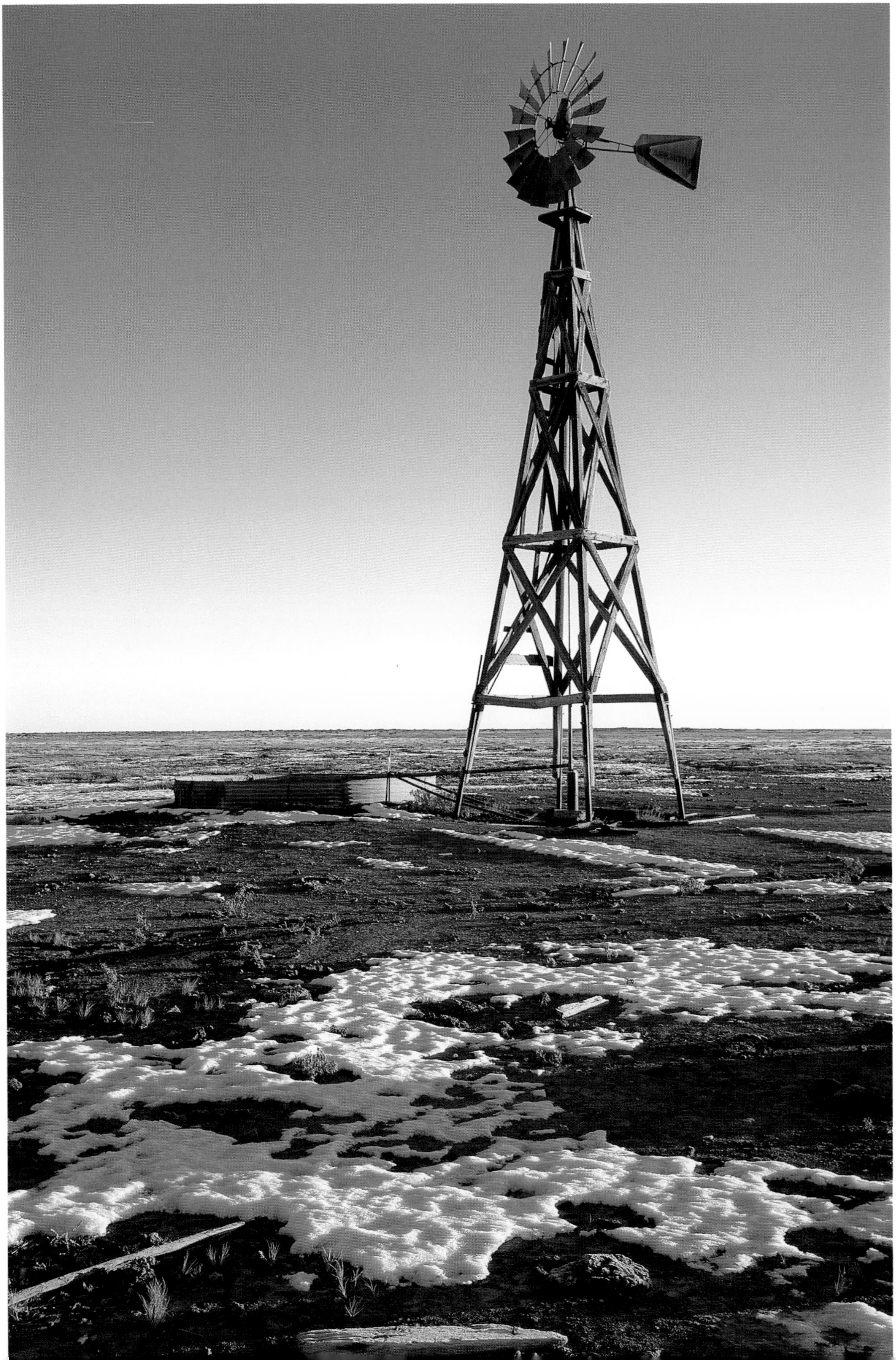

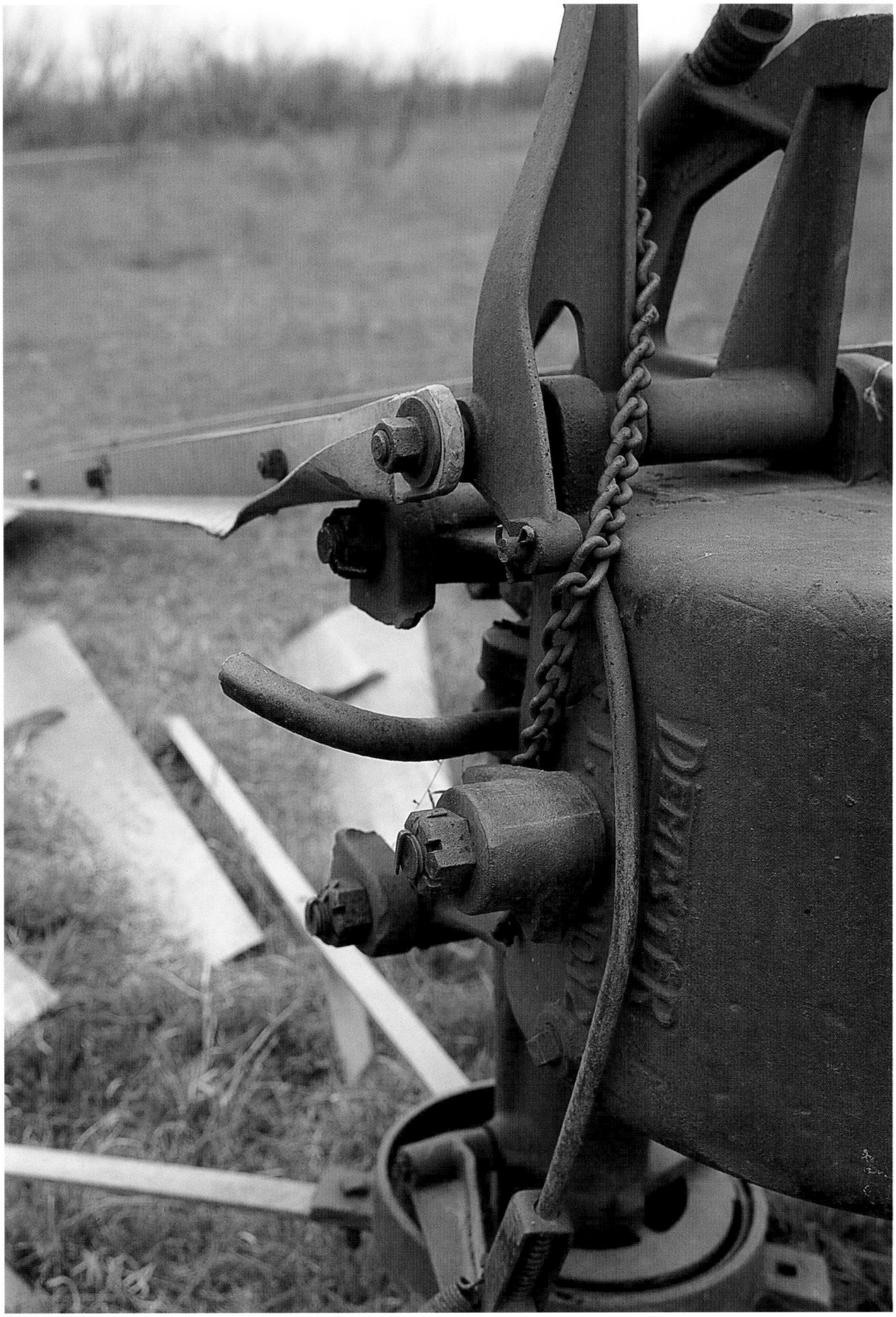

ONE MAN SAID THAT WHEN HE WAS A YOUNG MAN and before he left home to marry, his dad hired a fellow to repair the windmill. The family left for the day, and that evening when they got back his dad stared at the windmill and declared that something was wrong with it. The hired hand had already left, so Dad couldn't ask him about his work. Just then a breeze came up, and Dad said, "Will you look at that, it's not turning right." It seems the hired hand had put the wheel sections on backwards, which made the wheel turn the wrong way. Dad called the hired hand on the phone and made him come out the next day to "fix that left-handed windmill."

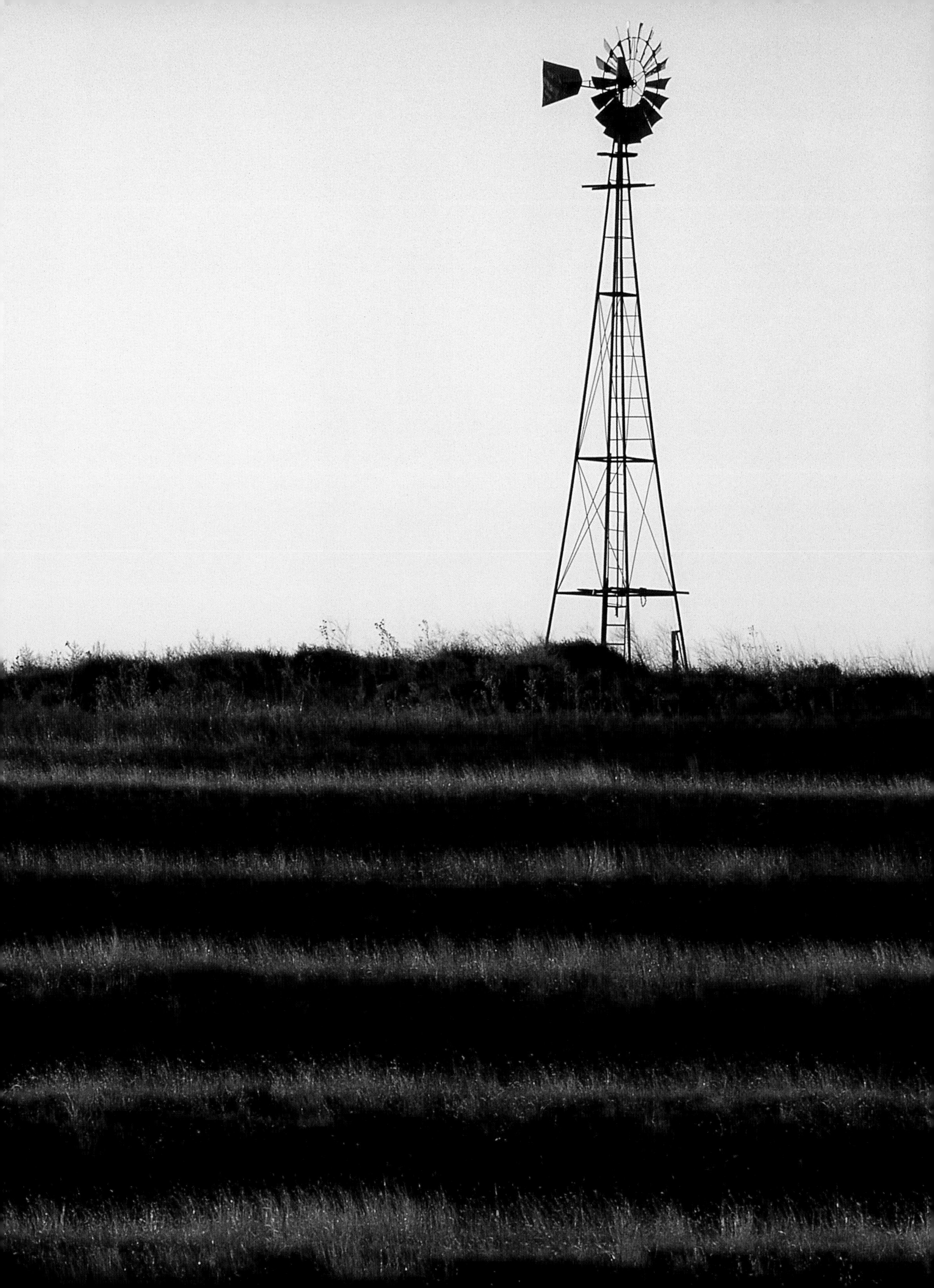

On a man's ranch in Central Texas, all the early Aermotor windmills were "open-gear" types. When the newer, enclosed-gear models became available after 1915, all the older, open-gear mills were replaced. His father always had the kids climb the tower to make those changes and repair the windmill. He wondered why his dad never did climb the tower and do the work himself. It wasn't until many years later, when he was grown, that he found out that his dad was afraid of heights.

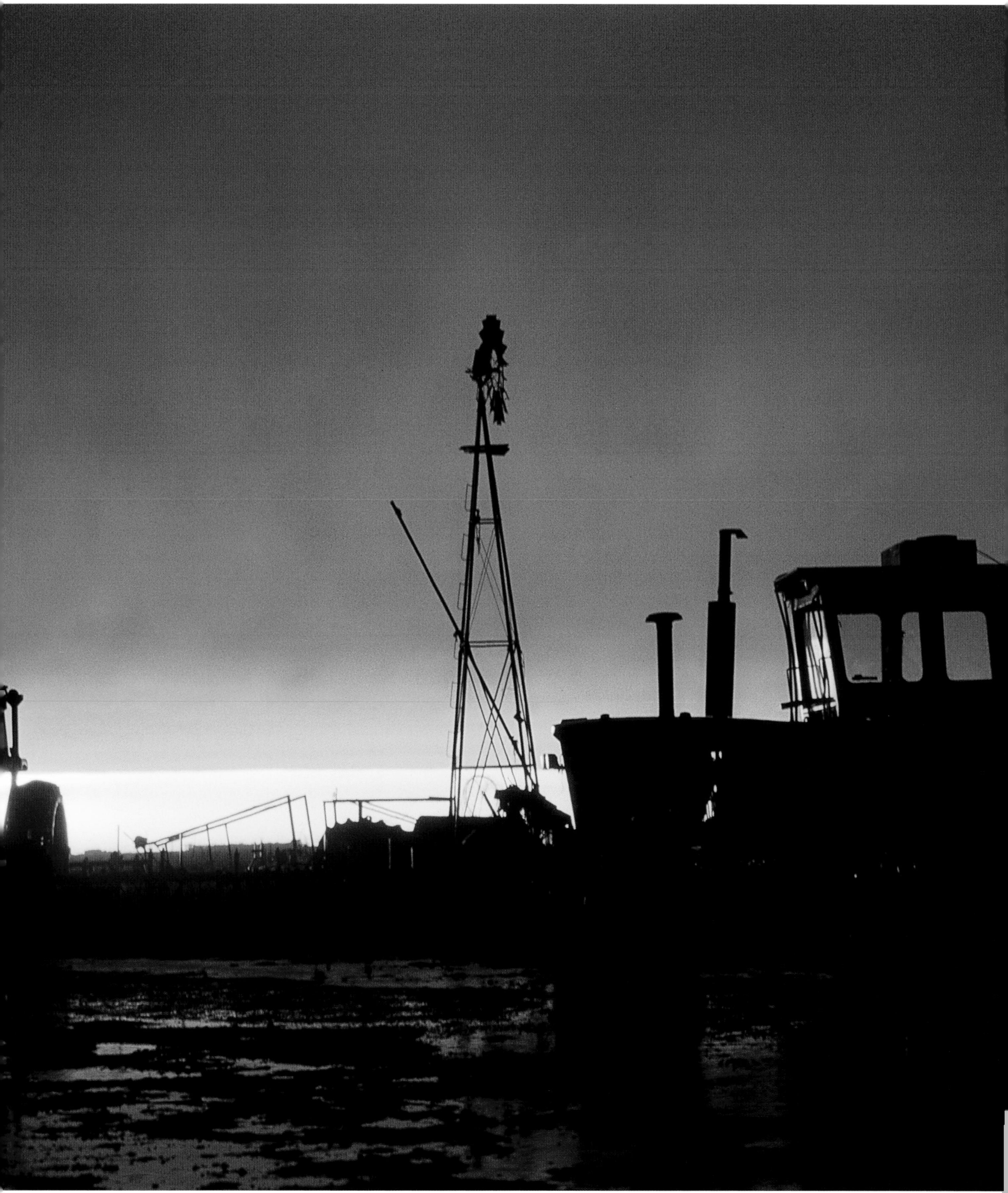

"In the early 1940s we lived on a farm near Idalou, Texas, where my first baby was born. There were always plenty of chores to do, and with the baby crying I would put him in his buggy and roll it out where he could watch the windmill turning. The windmill would usually entertain him until I could get the chores done."

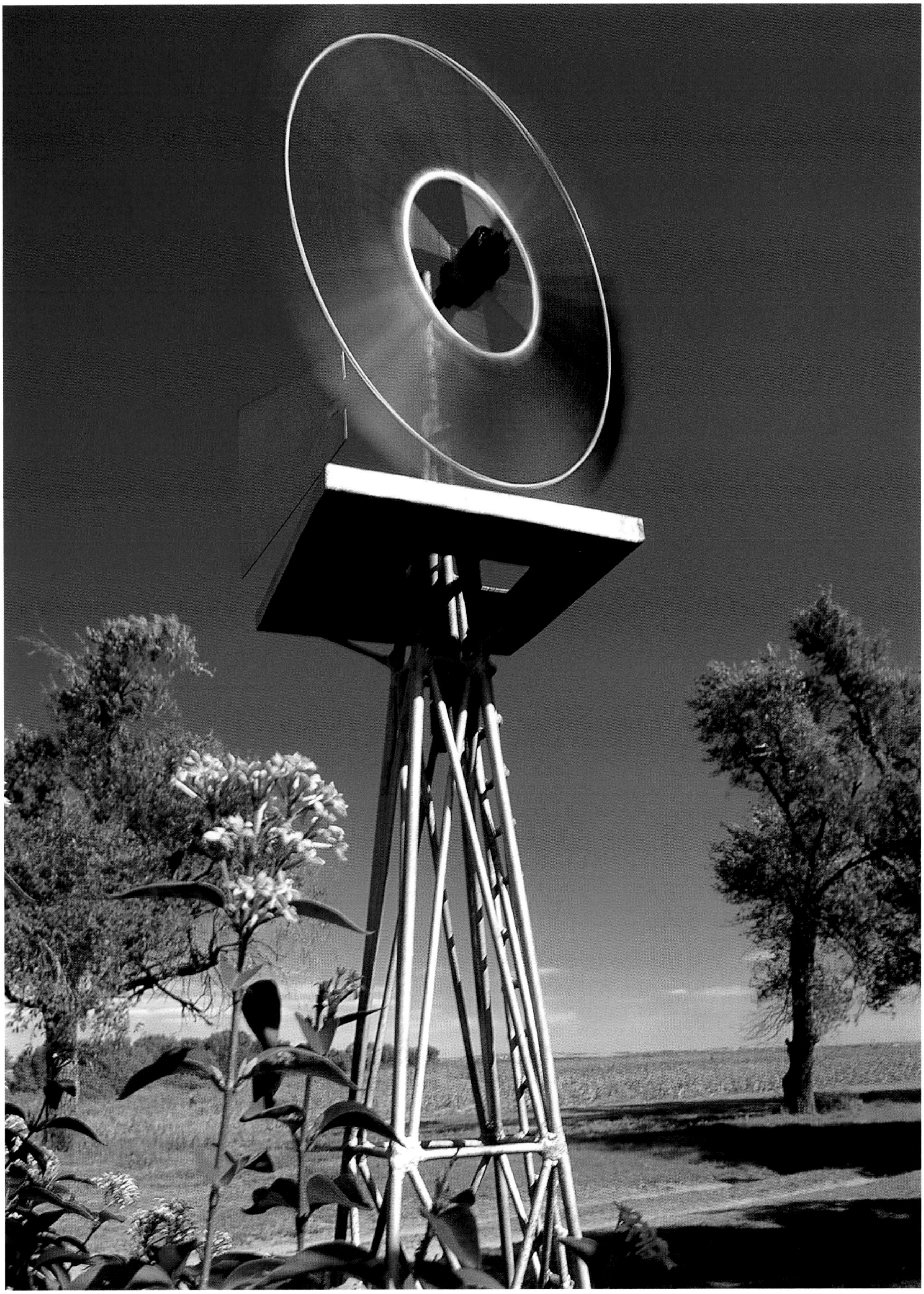

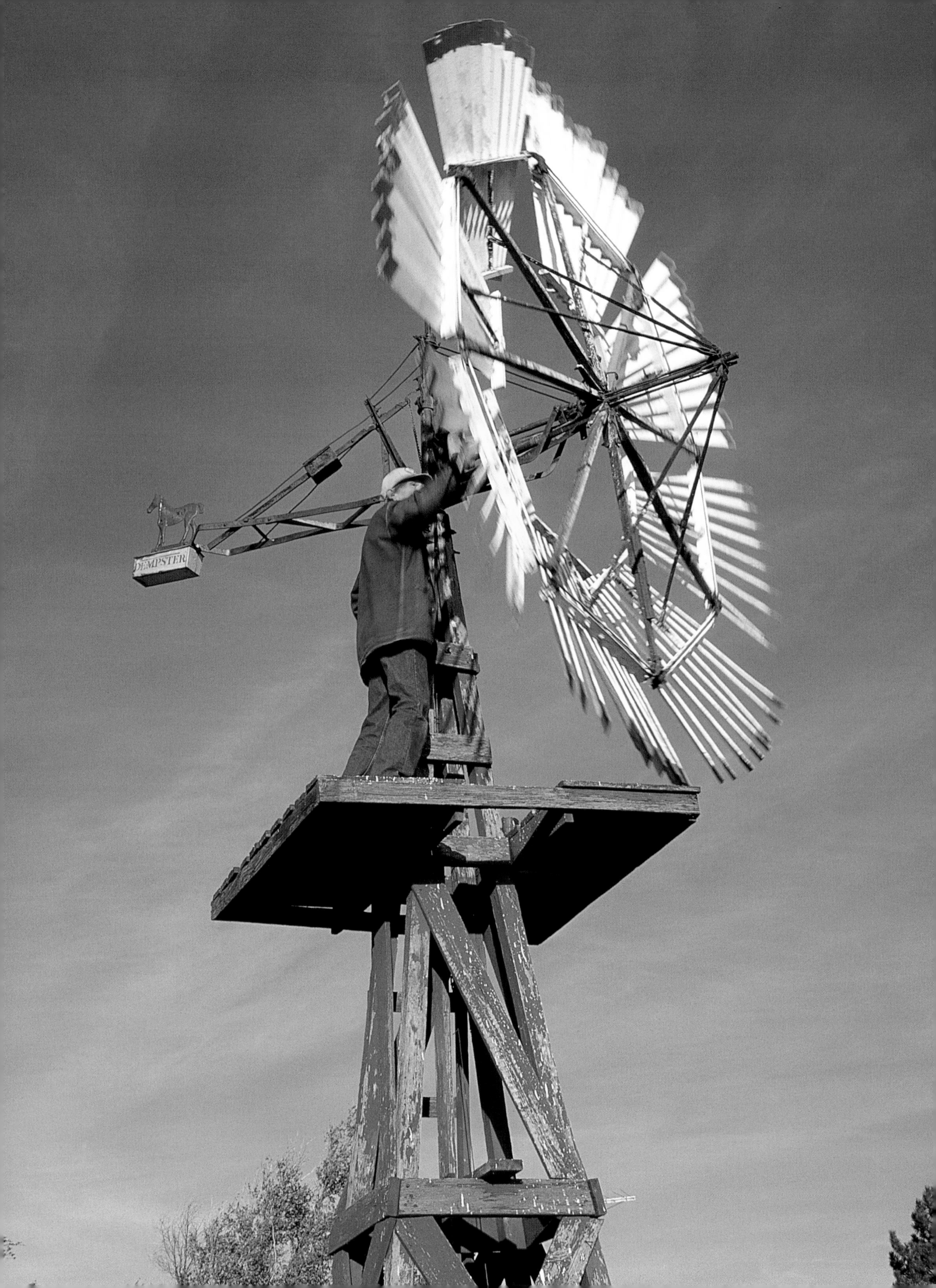
DEMPSTER

A SQUEAKY WINDMILL was the cause of a feud between two neighbors. Mr. Barker didn't oil his windmill as often as needed, and it bothered the Whitt family, especially at night. Early one morning Mr. Whitt awoke to see Barker climbing his windmill tower. He got out his shotgun and snuck close. As Barker reached the top, Whitt shot out the steps of the ladder. Barker turned around and yelled down, "What are you trying to do, kill me?" Whitt yelled back, "No, but I'm gonna leave you up there until you oil that windmill."

"WE LIVED IN THE COUNTRY and always had a windmill. Some friends of ours who lived in the city would come out to visit with their boys so they could have some 'country experiences.' One morning during one of their visits, I had washed my white chenille bedspread and hung it on the clothes line to dry. We missed one of the boys and went outside to check on him. Just as we got outside, the boy reached the top of the windmill and was getting ready to jump with my beautiful chenille sheet stretched out like a parachute. We stopped him in time, but my bedspread was never the same again."

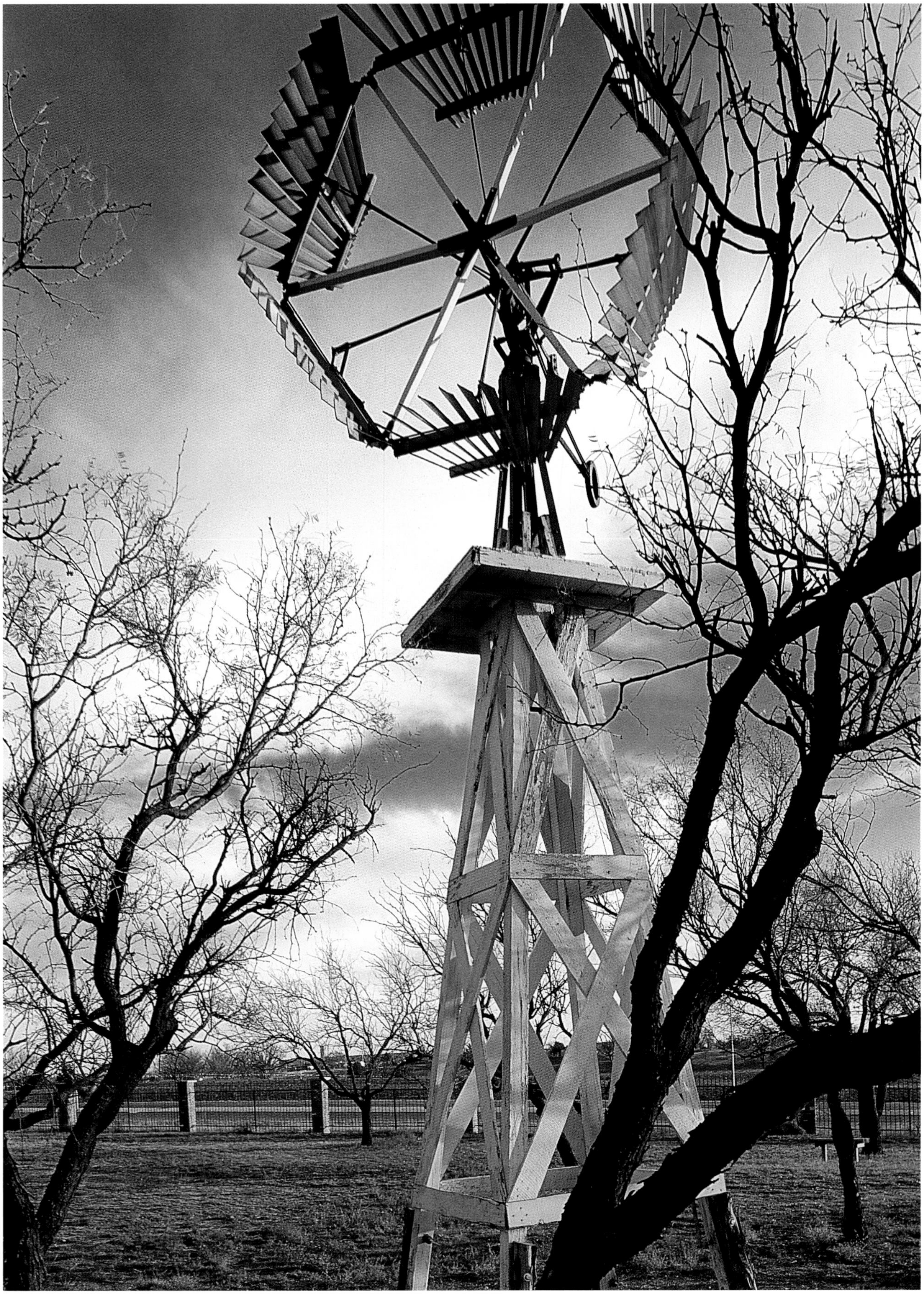

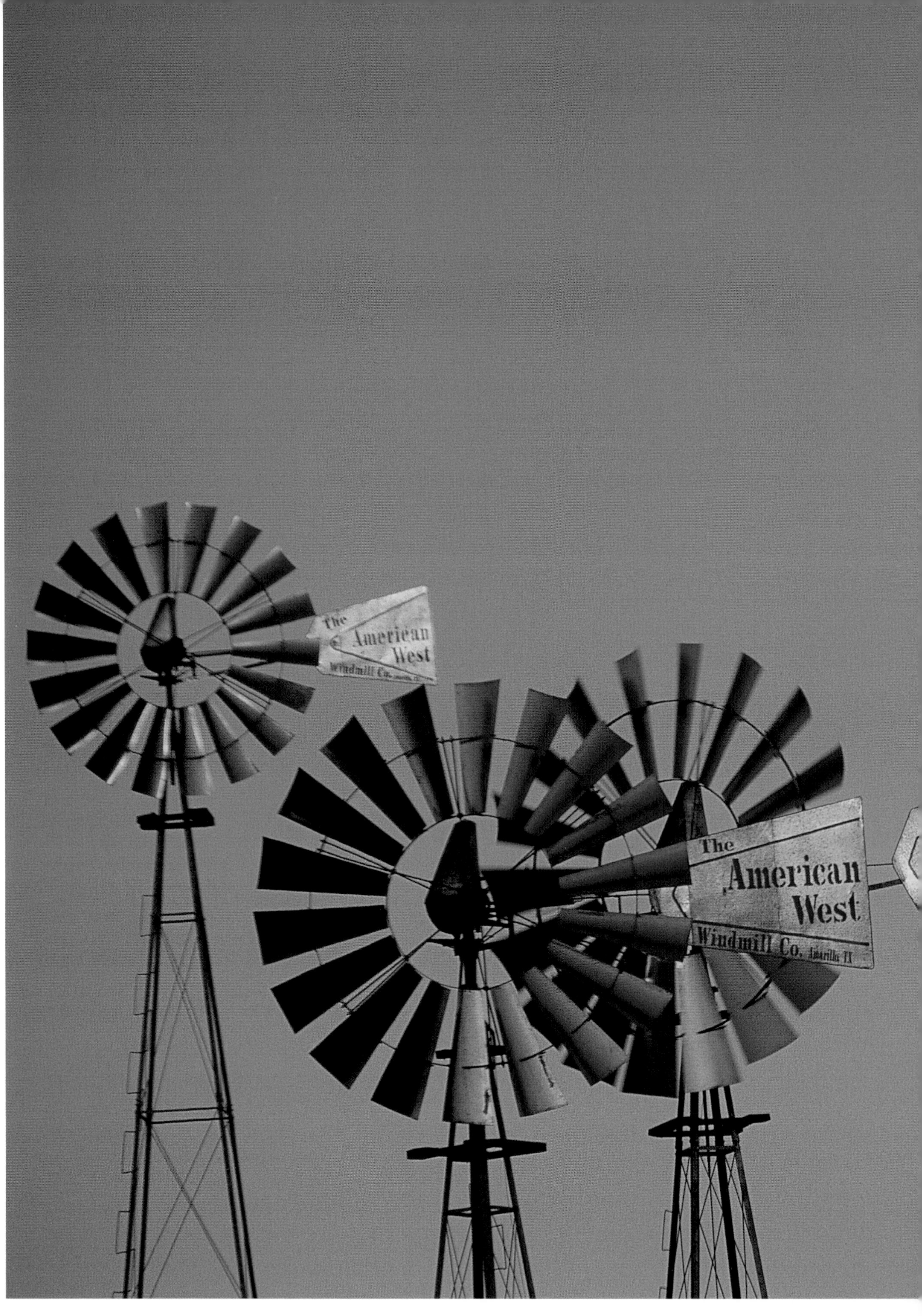
The
American
West
Windmill Co.
The
American
West
Windmill Co.

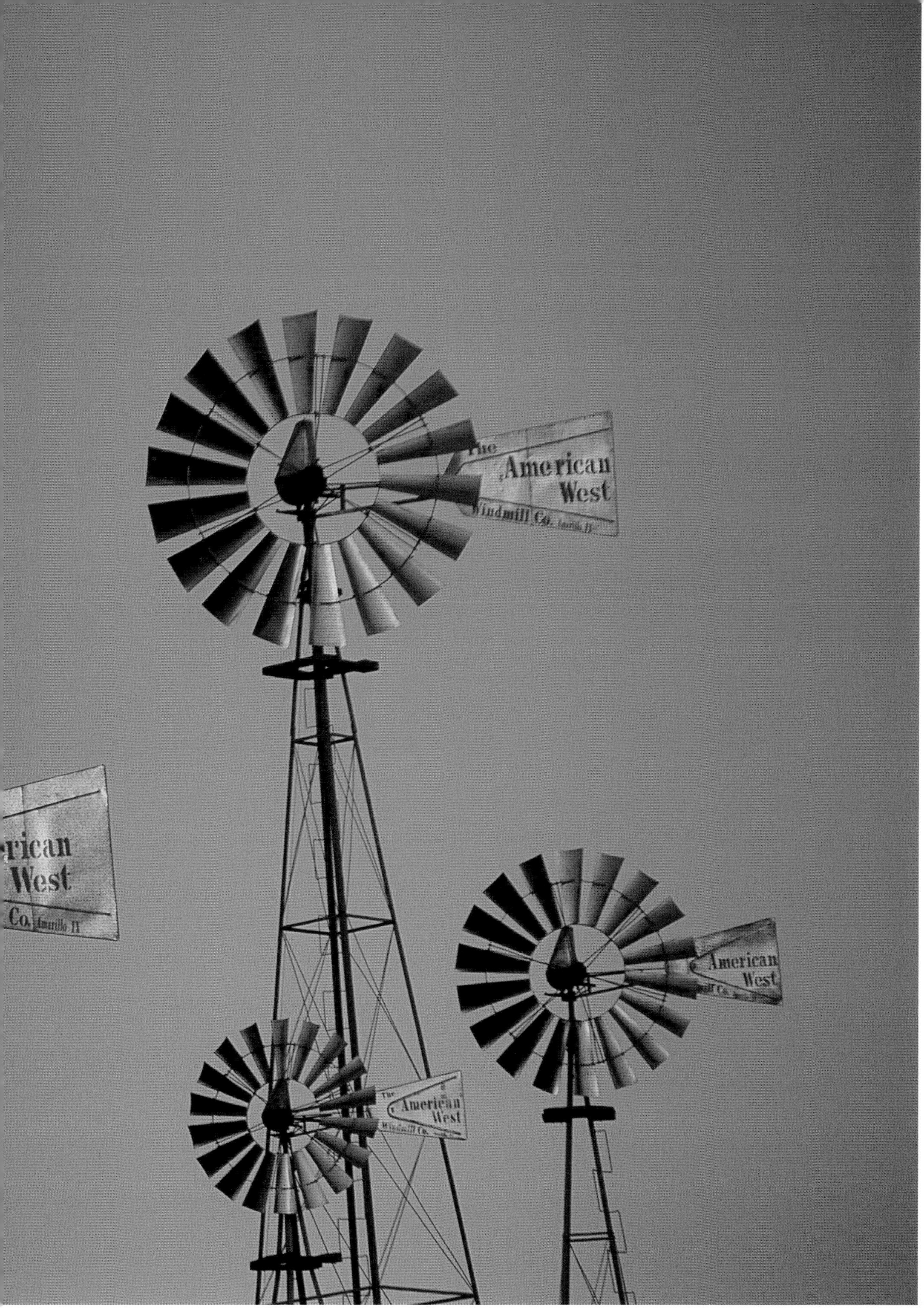
The American West Windmill Co.
rican West Co. Amarillo
American West
The American West Windmill Co.

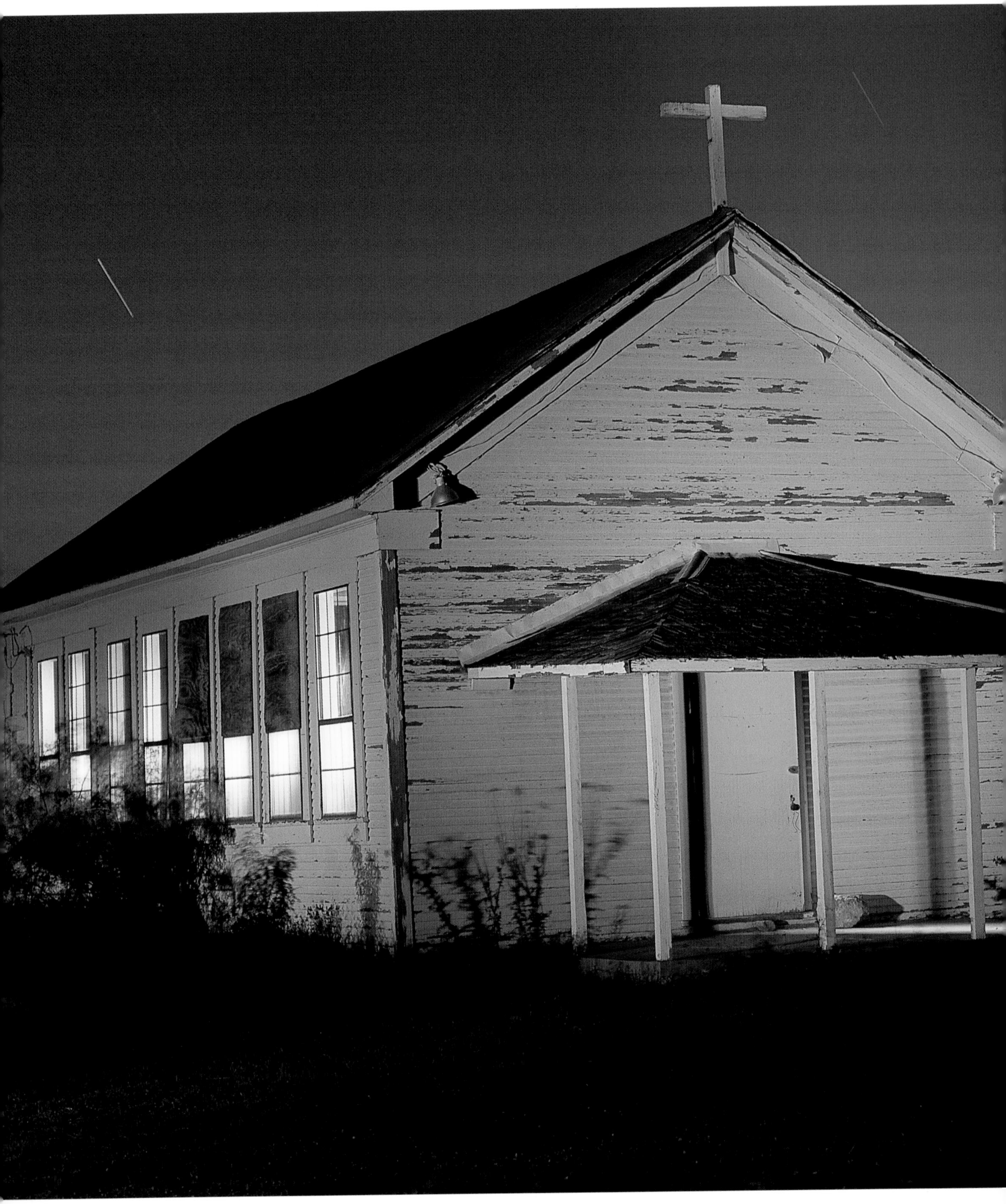

Johnny Williams, a retired Methodist minister, used to live in Earth, Texas, where the Parish windmill was made. The phone company had installed new lines, and everyone got a new number. Williams just happened to get the old phone number of the Parish windmill manufacturer. In the middle of the night he got a phone call, but, being a preacher, he was used to late night calls and went ahead and answered it. The man on the other end told him that he had to get out there and work on his windmill. Mr. Williams said, "I don't think you understand; the phone number has been changed, and I am not who you want." But the man was insistent that he had the right number and wanted him out there now to work on his mill. Mr. Williams said finally, "Well the best that I can do is pray for it."

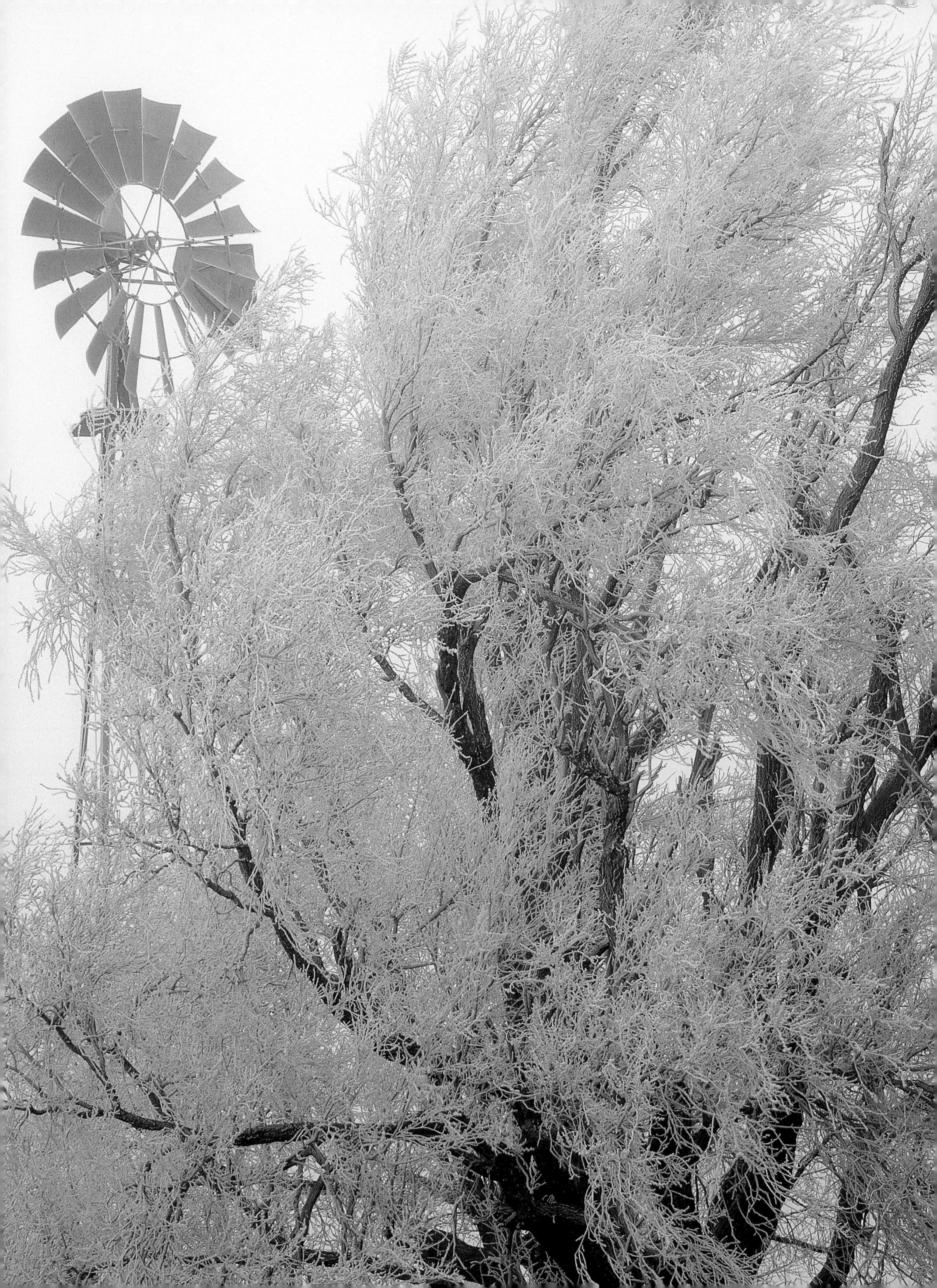

AN IOWA FARM COUPLE HAD A WINDMILL that needed some grease. As Mr. Day climbed up the tower with a bucket of grease, he saw his wife heading out to turn on the windmill. He was already at the top of the tower when he yelled, "Don't turn on the mill!" as loudly as he could. She was hard of hearing, however, didn't hear his yelling, and turned on the windmill. The tail swung around and promptly knocked him to the ground. He did recover, however.

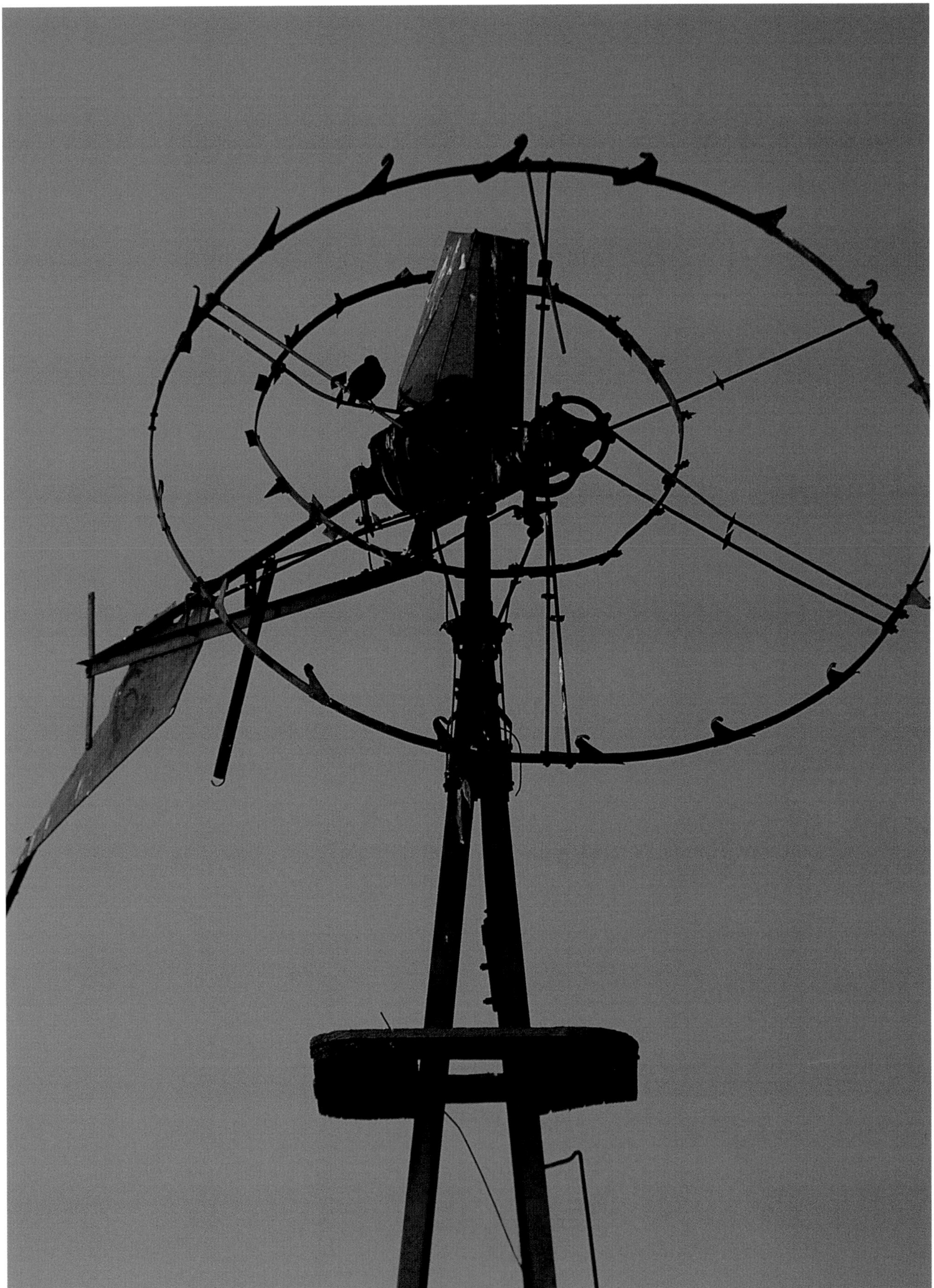

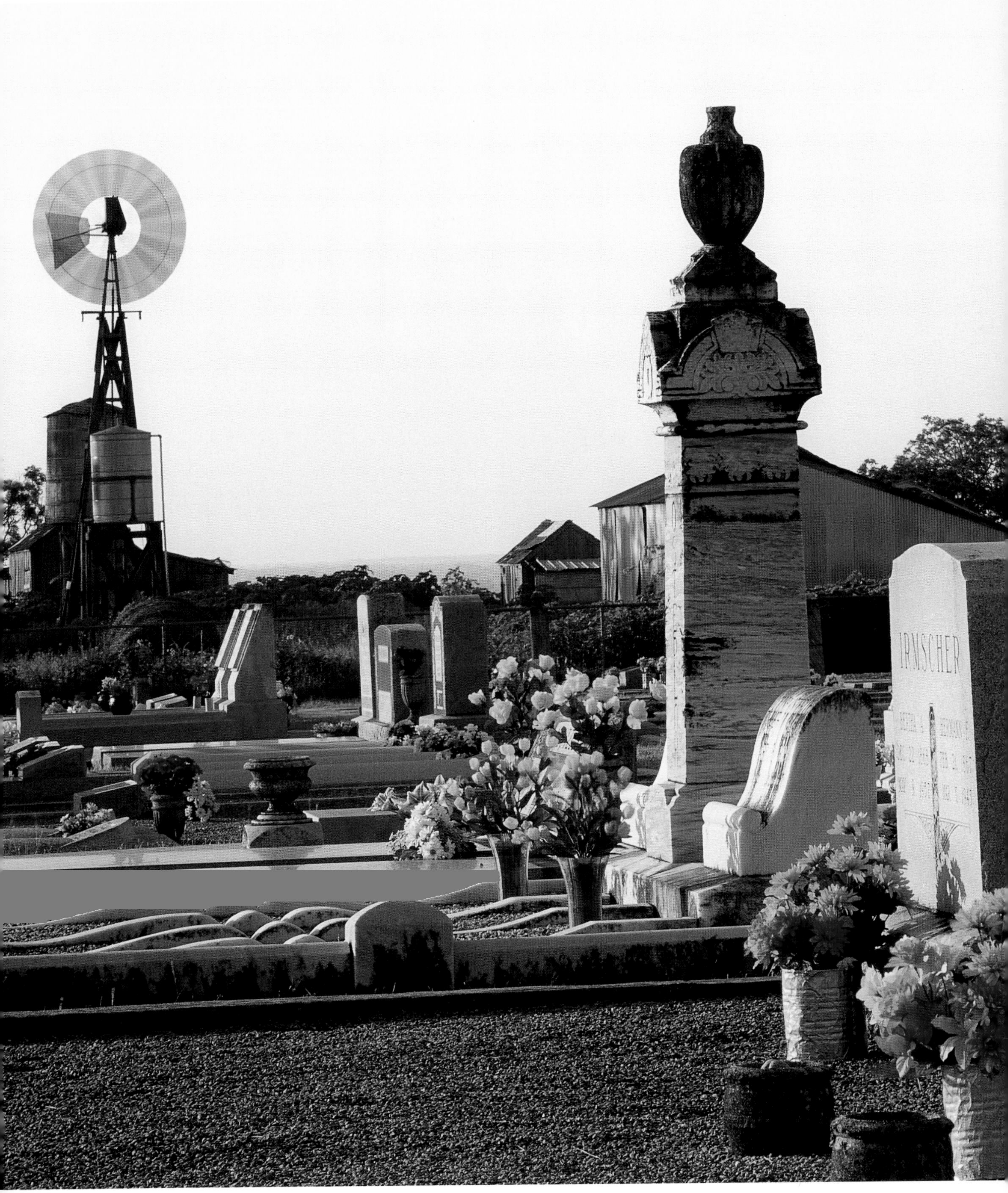
IRMSCHER

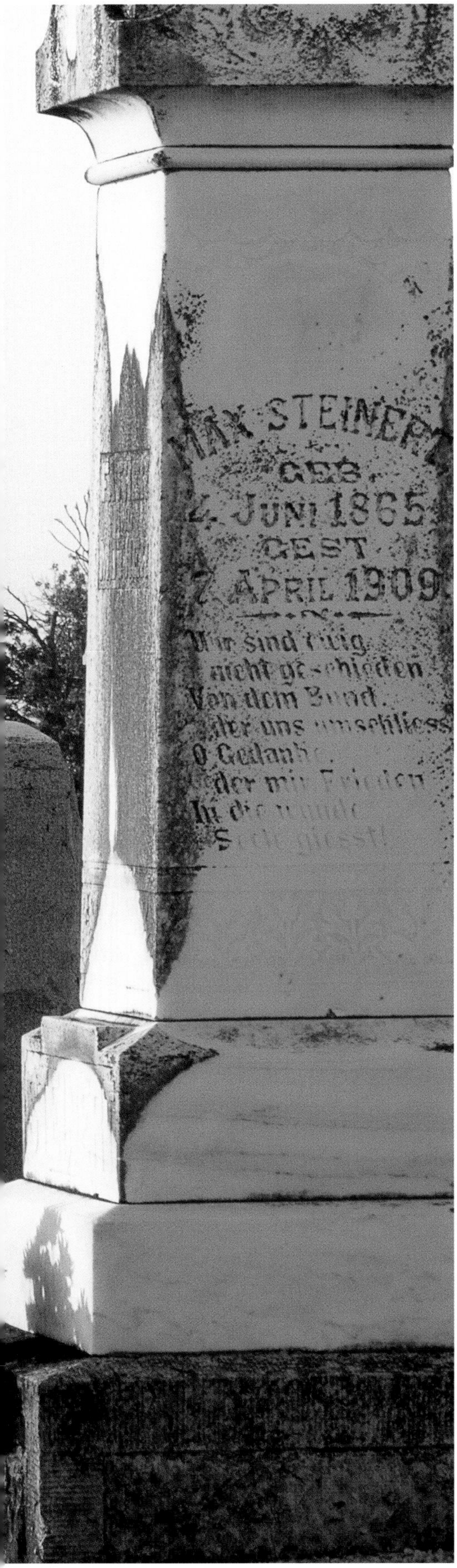
GEB.
JUNI 1865
GEST
APRIL 1909

"OUR FARM HAD A CEDAR WATER TANK that was elevated about fifteen feet to provide running water into the house. It was one of the few houses around that had pressured water. A neighbor had a death in the family, and since it was a hot summer day, my parents asked their family if they would like to take baths before the funeral. As soon as they began their baths, the wind died down and the mill stopped pumping. Only a couple of family members got to take their 'funeral' baths before they ran out of water."

MARTIN MOORE WORKED AS FOREMAN on a ranch near Post, Texas. The owners didn't visit often, usually only twice a year. Before one such trip he had a chance to buy a used windmill that was in perfect shape and at a very low price. It was put up and operated perfectly. When the owners arrived, he decided to take them out to see what a great buy he had made. As they drove along the ranch road he couldn't seem to see the windmill, but he knew its exact location. When they reached the site, they saw the windmill and tower lying on the ground all broken, twisted, and bent out of shape. It had been hit by a tornado. He said he felt sick, because the owners never got to see what a bargain it had been.

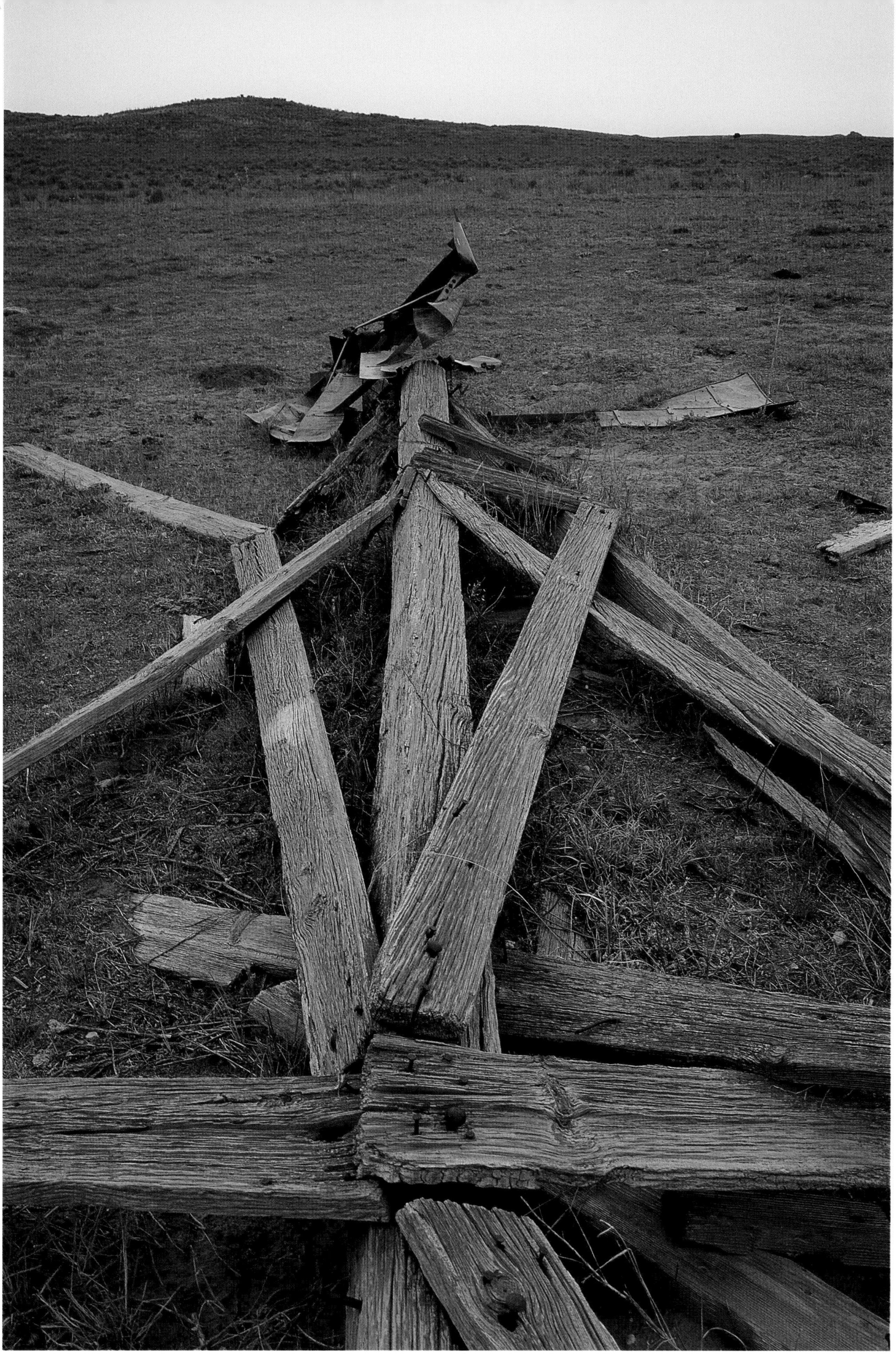

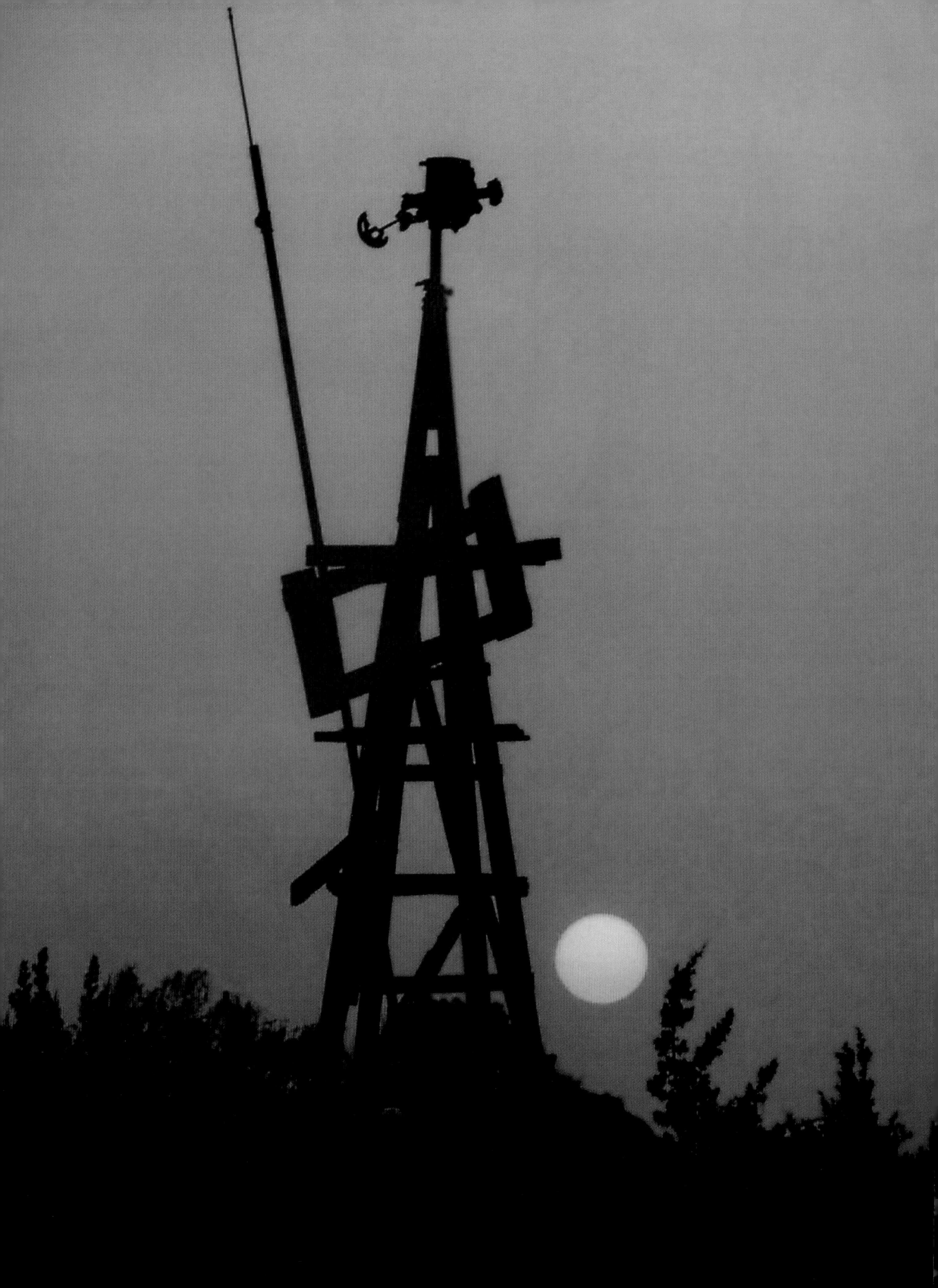

A SEVERE HURRICANE HIT THE GULF COAST in the 1960s. One woman tried to ride out the storm, but the house she was in became flooded and eventually collapsed. All she could do was keep praying that she would live through it. When the storm was over, she came out of her crumpled house to see complete devastation. She looked up into the sky and saw a cross, which she followed to another house that had survived. There was a family in the house, and they offered to help her. The cross was a windmill that had survived the storm, still standing, but with only two perpendicular wheel arms left on the hub.

ONE OF THE FAVORITE THINGS TO DO on a windmill when you were a kid was "to climb the windmill tower, get ahold of the tail, run around the platform, and jump, holding on to the tail. The tail would go swinging around with me hanging on for dear life." When asked if it was scary, we heard, "Only the first time you do it."

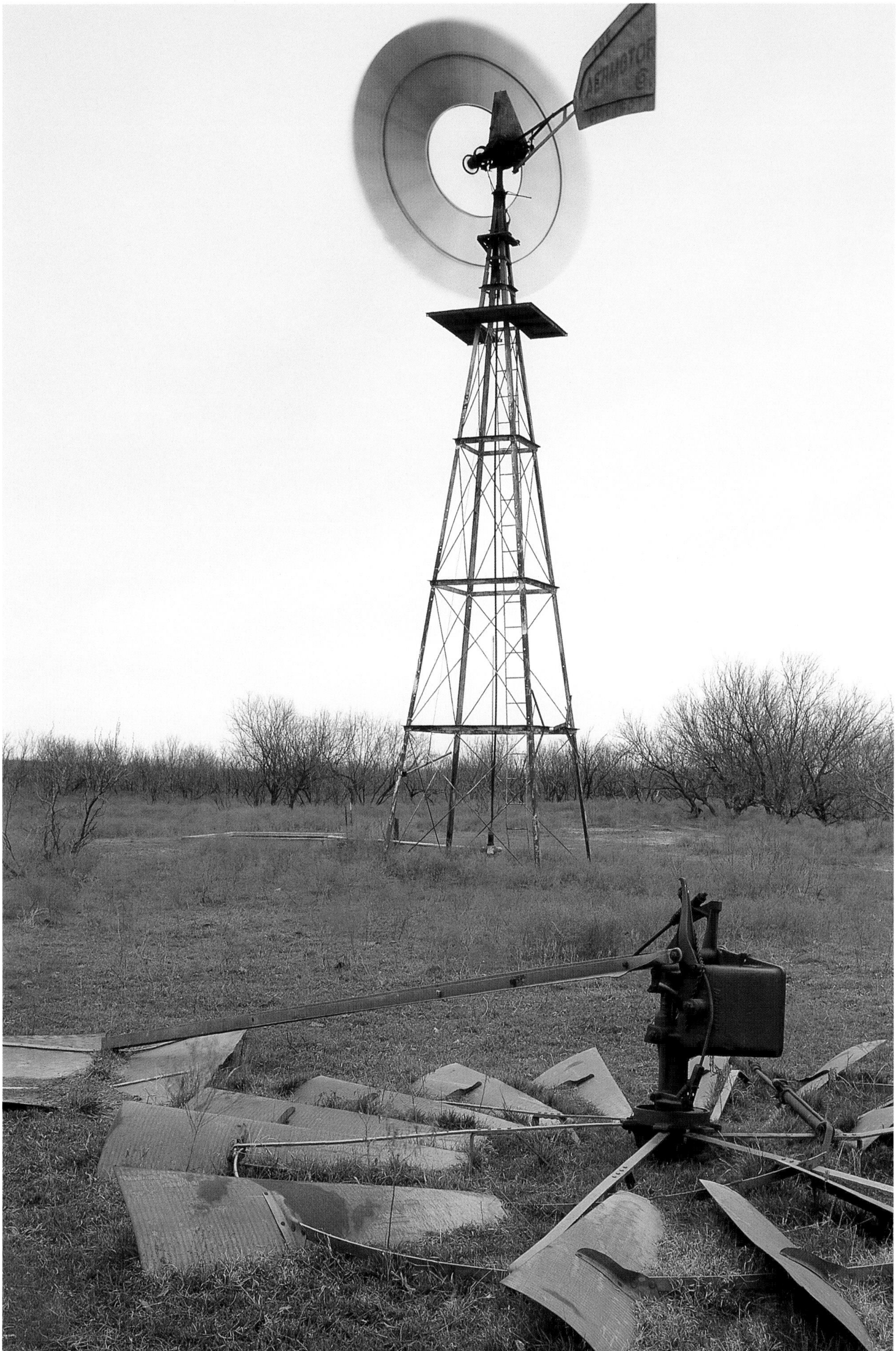

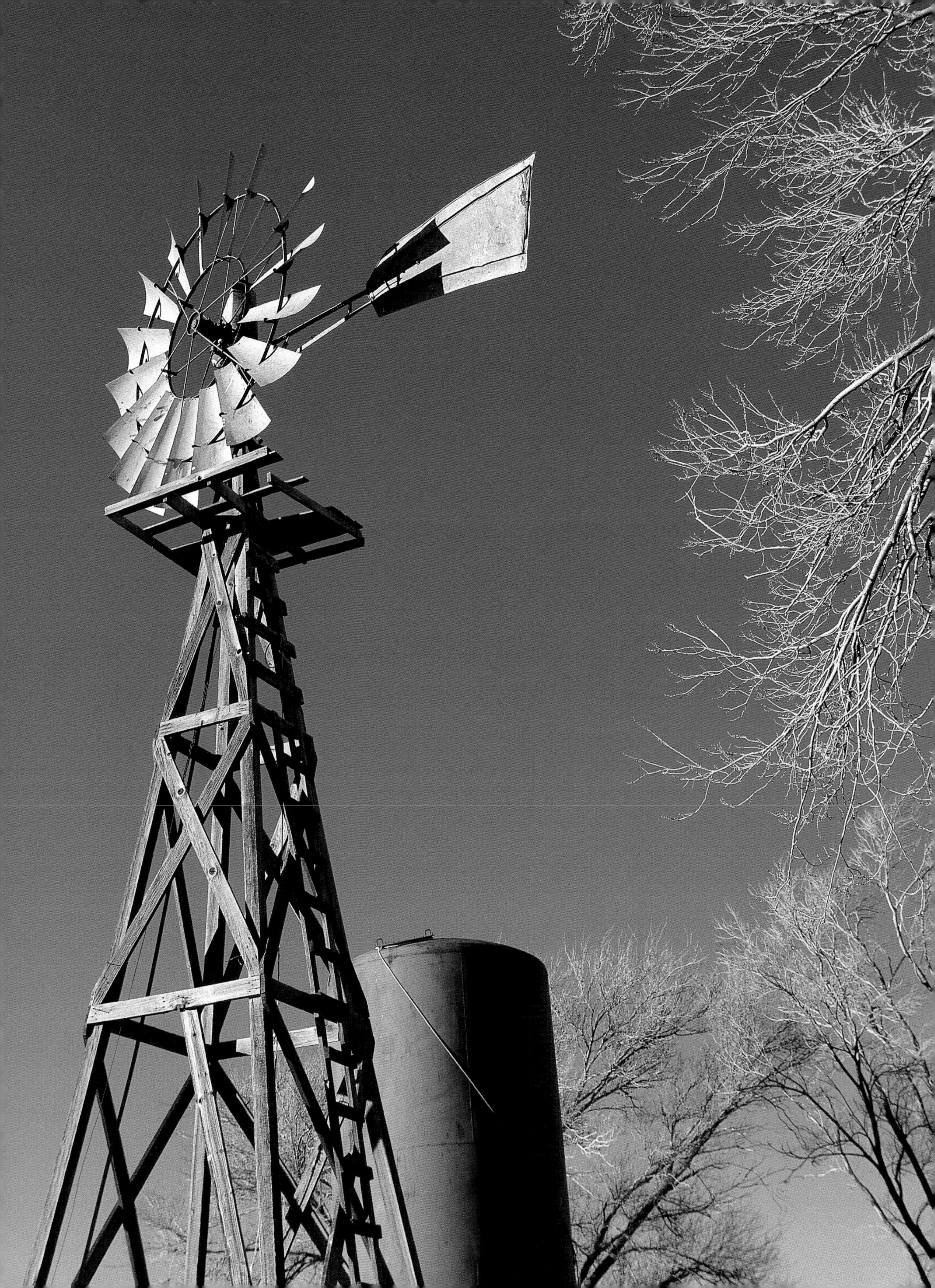

WHEN RICK NIDEY WAS GROWING UP, his dad secured their rickety old windmill with guy wires. One of the wires was tied to a propane tank on the ground not far from the windmill tower. When Rick was about eleven, he and a friend decided to slide down the guy wire with a hook. As the boys would slide down, they put their feet down on a mound of dirt that was over the well house and ran the last few feet so that they could stop before hitting the tank. One day a friend's little brother wanted to try it, so they helped him up to the top of the tower. As he started to slide down they realized that he would not be tall enough to put his feet down at the dirt mound to slow down. He came sliding down past the well house mound and didn't stop until he hit—full speed—the propane tank. Rick said that he can still hear the "whonk" of that kid hitting the tank and "nearly knocking himself plumb out."

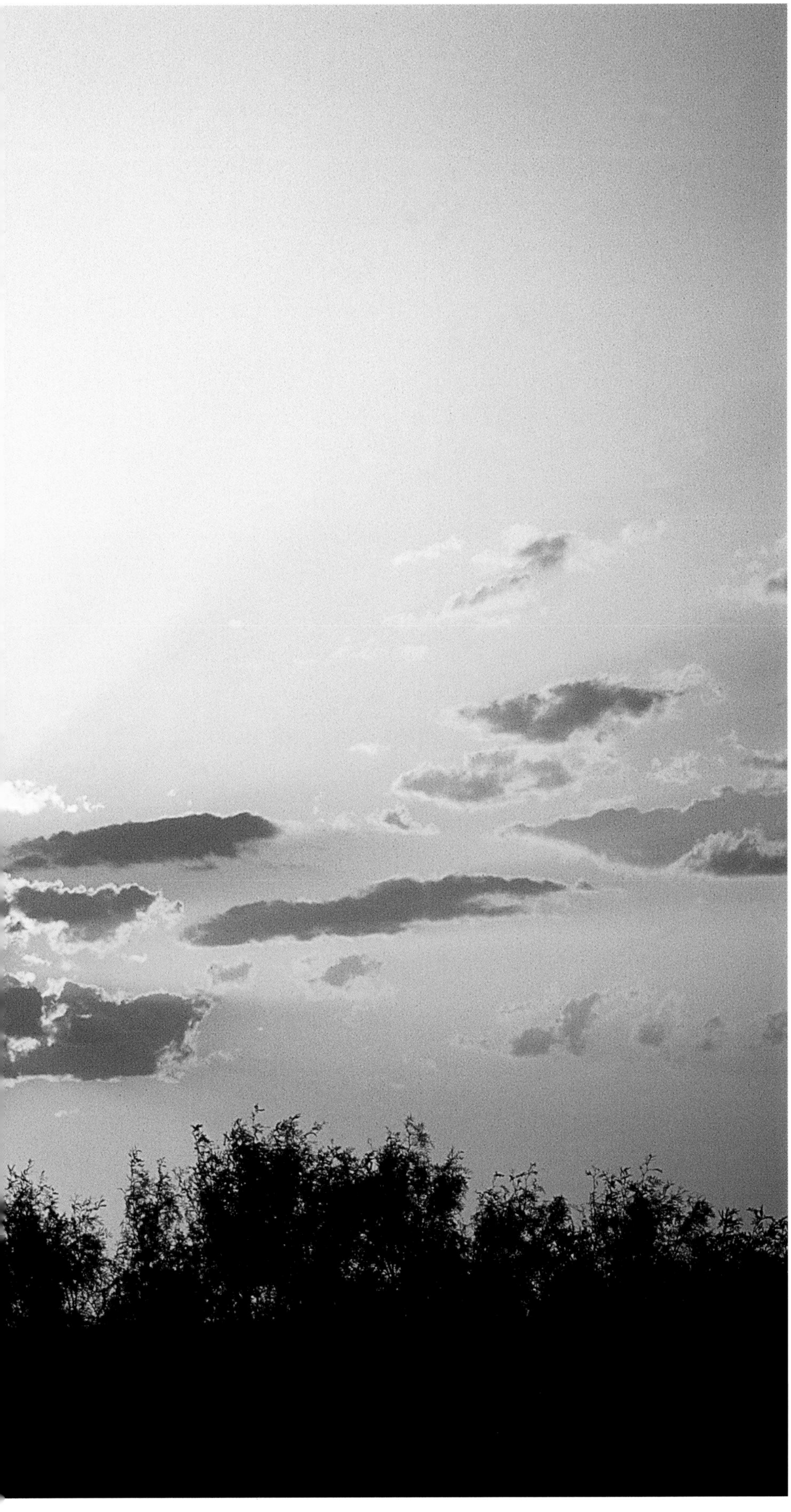

MRS. SMITH'S FAMILY MOVED to Cochran County, Texas, in 1922. Among their livestock were some old mules used to work the land. She said the mules were bad about running off, and her father would get up on the windmill with a telescope to find out where the mules had gone. When he spotted them, he would get on his horse and go round them up.

A VISITOR EXPLAINED THAT HER MOM AND DAD were early settlers on the South Plains in the late 1800s. There were very few towns in those days, and the farm where they lived had no close neighbors. Her dad worked many days from dawn to dusk and far enough from the house that her mother was pretty much alone. She was afraid to be by herself, as there were still some Indians around. As soon as her dad would leave in the morning, her mom would climb to the top of the windmill tower and stay there until she could see her husband heading home. She continued to do this every day that he was gone, even up to her first child. She took the baby with her and then later his little sister. By the time she had the third one, she said it was just "too darn hard to get three little kids up there."

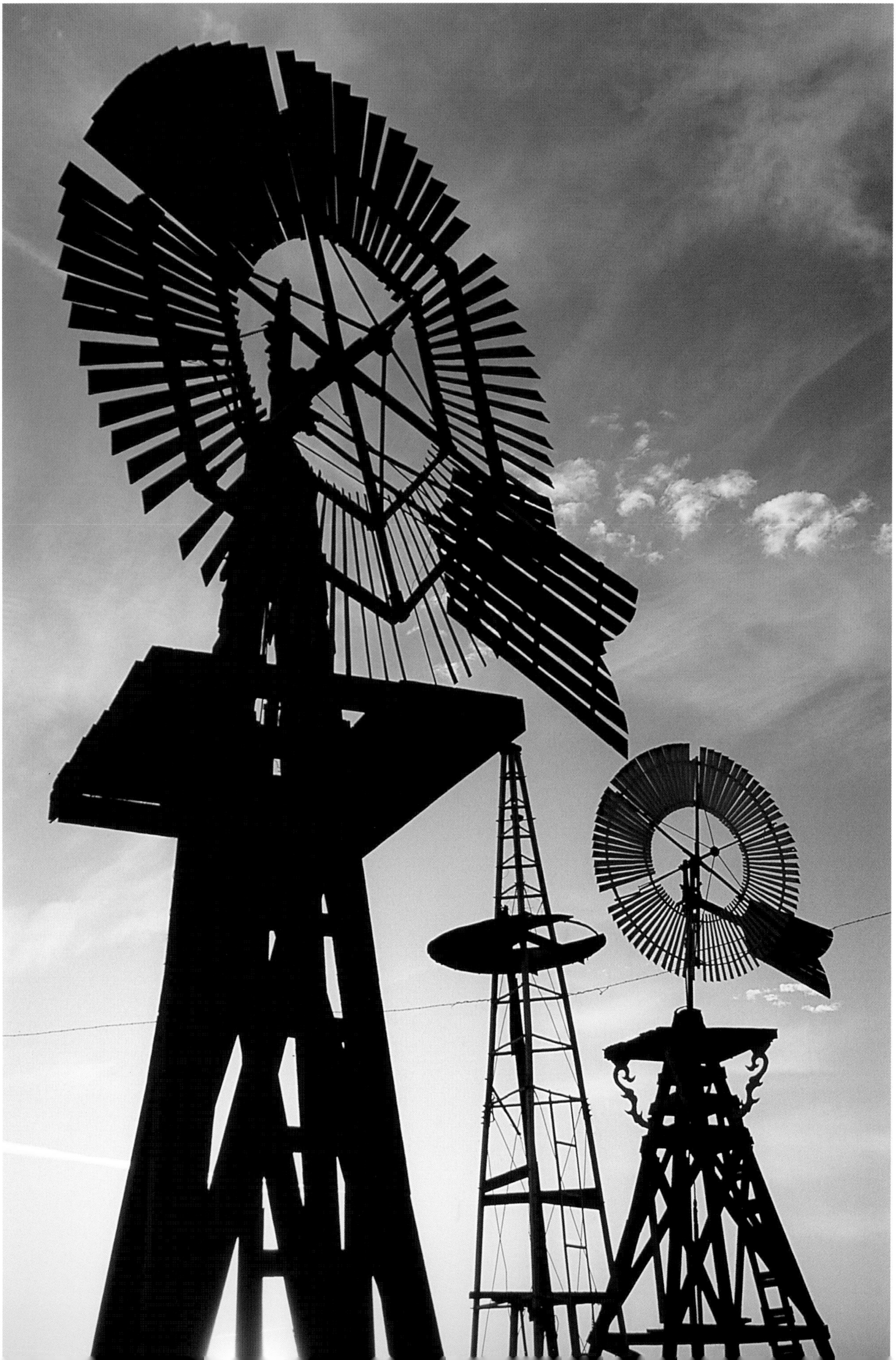

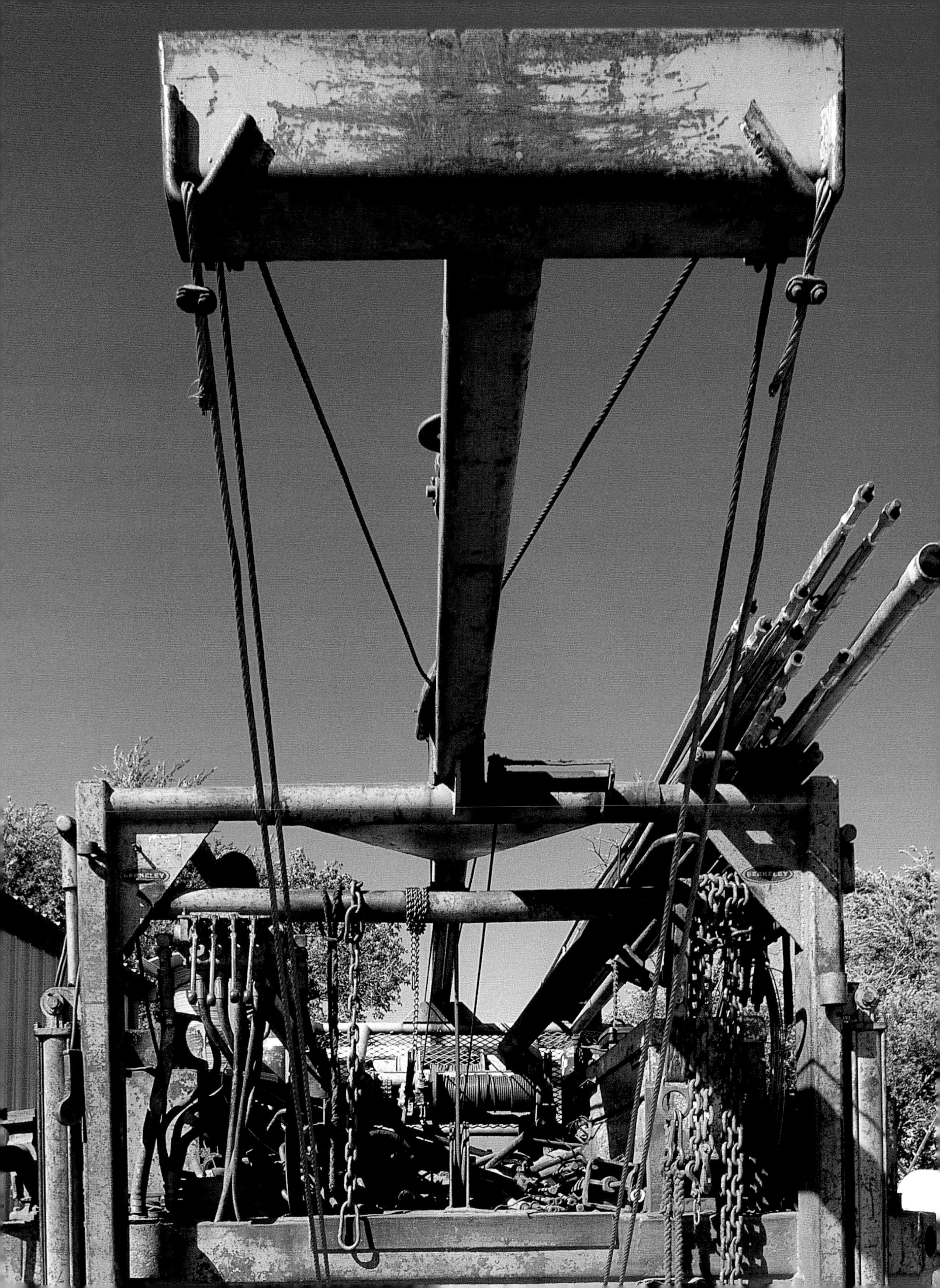

AN OLD WELL DRILLER RETOLD this great windmill story: "When I was drilling a very deep hole, I took along a young hired hand named George. This boy had helped me around the shop but had never worked on a drilling rig before. I had bought a new drilling truck and had lined up a job on a ranch in Oklahoma. George was trying too hard and dropped tools and stumbled over things as we set up the truck. We started drilling and got the well down about two hundred feet. We had just pulled the bit from the hole for an inspection, and I went around the truck to get a drink. From the other side I heard George say, 'Uh-oh!' and I walked back around to where he was standing. He was looking desperately down the hole. 'Dropped my pen knife,' he said. It was too small an item to worry about, so we kept drilling.

When we got down to three hundred feet, we pulled the bit again to change it. While we were unscrewing it, George made a funny noise and I looked at him and yelled, 'What did you drop now?' He said, 'I guess it was that screwdriver I had in my other pocket.' I told him to take everything he had out of his pockets and everything from his belt, and while he was doing that, gosh dang it if he didn't drop a wrench down the well, too. I was jumping up and down, and George was walking around the well with his head down and a real sorry look on his face when he kicked two rocks and a small piece of chain down the well. I really got excited then, and with all those tools and stuff down the well, we just moved the entire rig to a new spot and started a fresh hole. I made George sit in the cab the rest of the day and wouldn't even let him out to go to the bathroom."

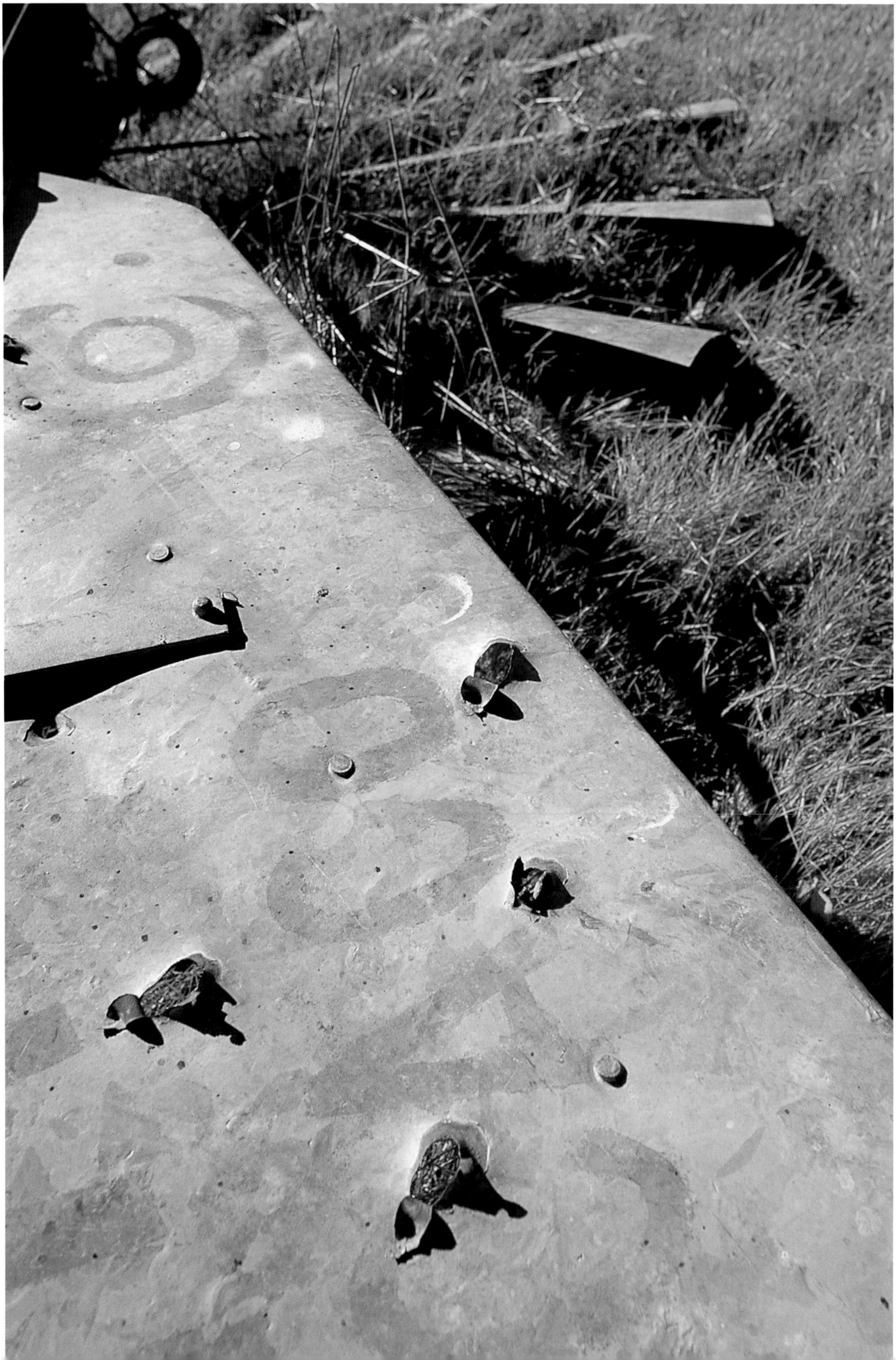

ONE OF RICK NIDEY'S BUDDIES had just put up a new windmill when he heard the "ping ping ping" sound of someone shooting at the tail. He looked around and saw a couple of kids in a new pickup with a rifle. He went up to them and said, "Why, that is a fine looking rifle you got there. Mind if I have a look at it?" They handed him the gun, and while he was inspecting it, he took serious aim at their new truck and pumped six shots right into it. As he observed their shocked looks, he said, "There, that will teach you not to shoot up other people's new windmills."

"IT HAD BEEN A DRY SUMMER, so we sold off the cattle and went across state for a week. We had some horses left in a pasture and asked our neighbor, a much older man, to look out for them and see that they got water every day. When we got back to the ranch we learned that the old man had died four or five days earlier. We just knew that all our horses would be dead, and we set off to look for them. Sure 'nough, we noticed all the windmills that we passed had stopped turning, and their stock tanks were dry. In the pasture where the horses had been fenced, though, was a Parish windmill, slowly turning in almost no wind, and the horses were standing around the stock tank. It seems that that Parish windmill had turned just enough to keep water in the tank so the horses could drink."

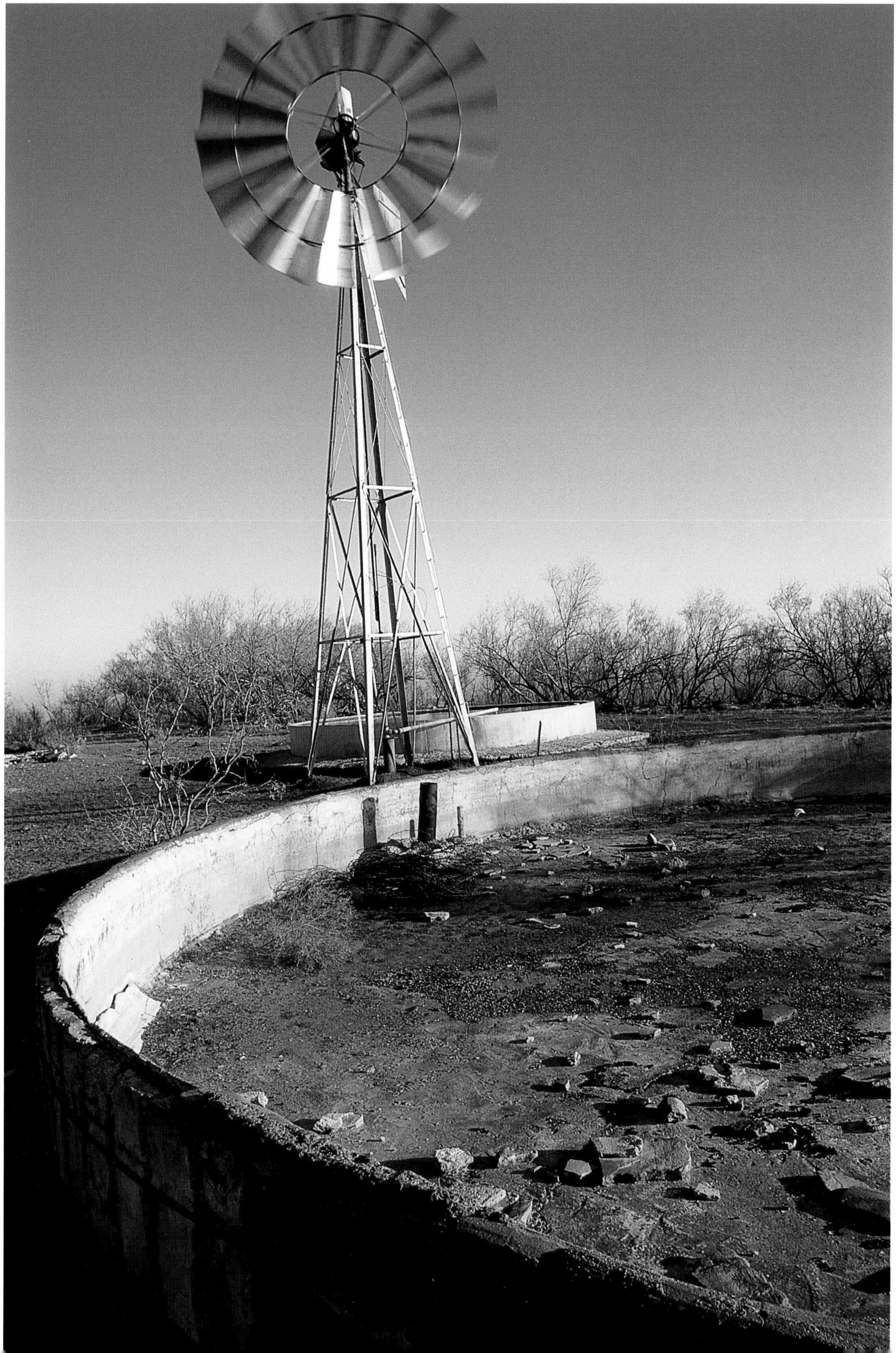

Coming 'Round the Corner

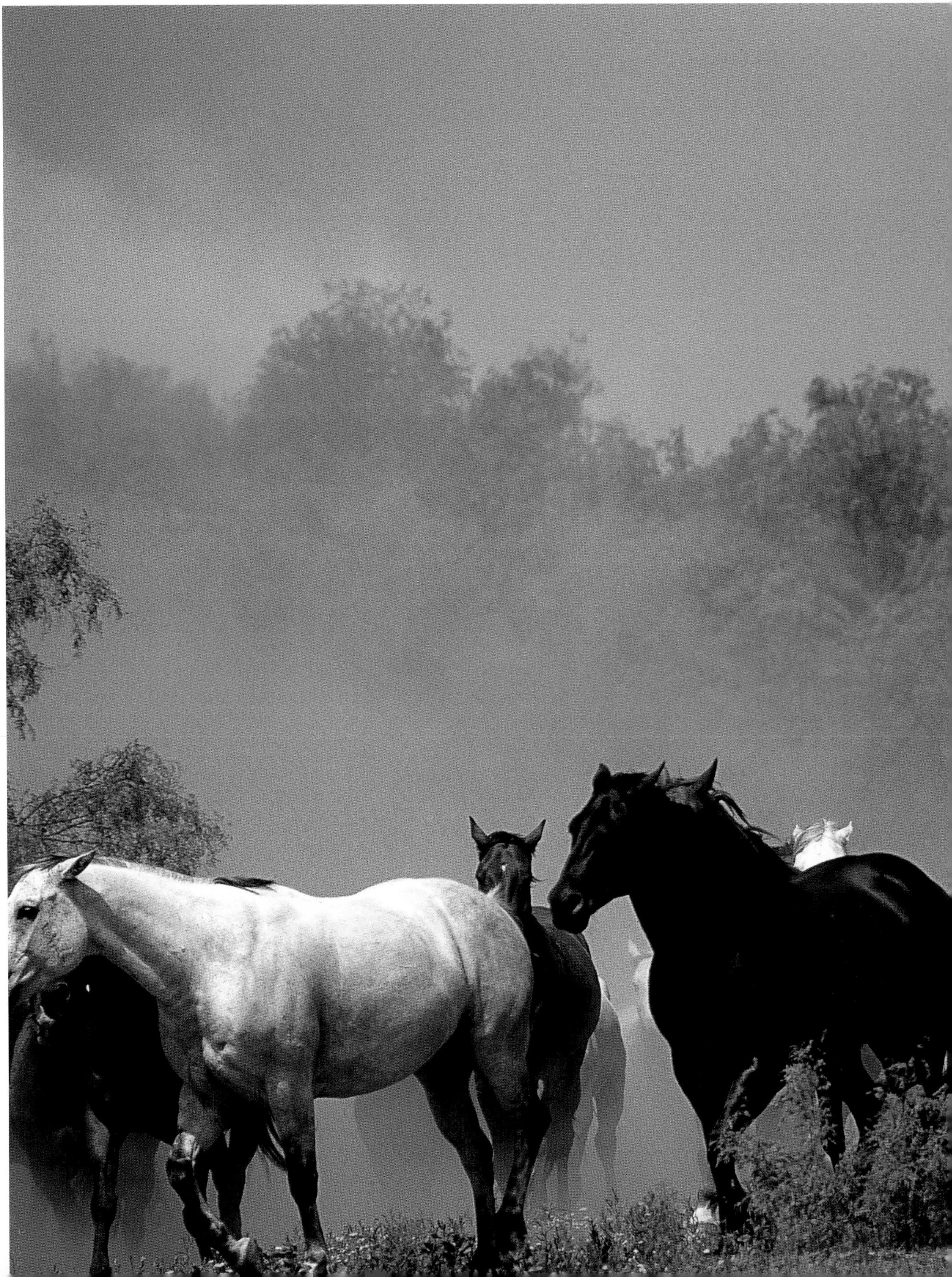

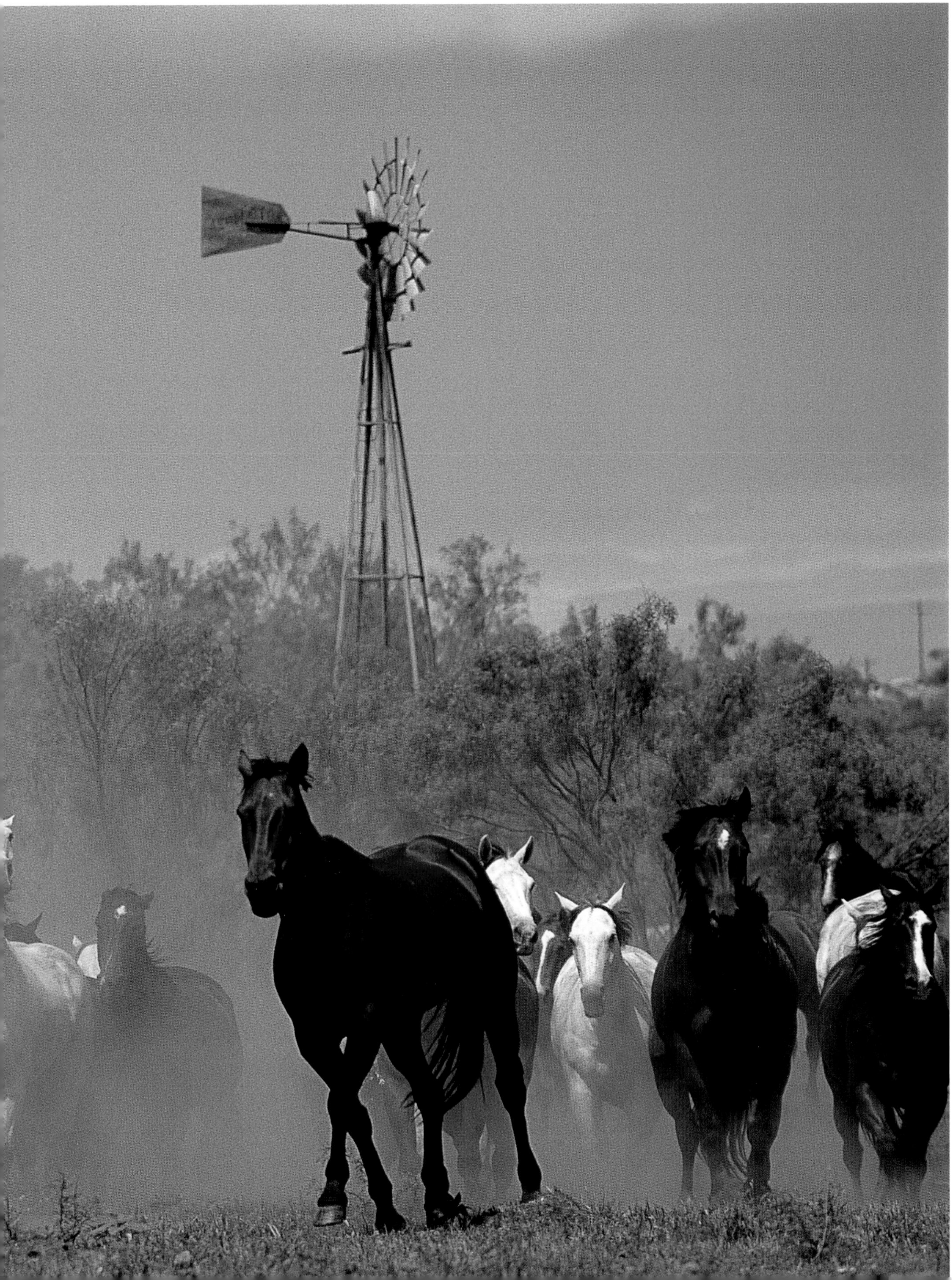

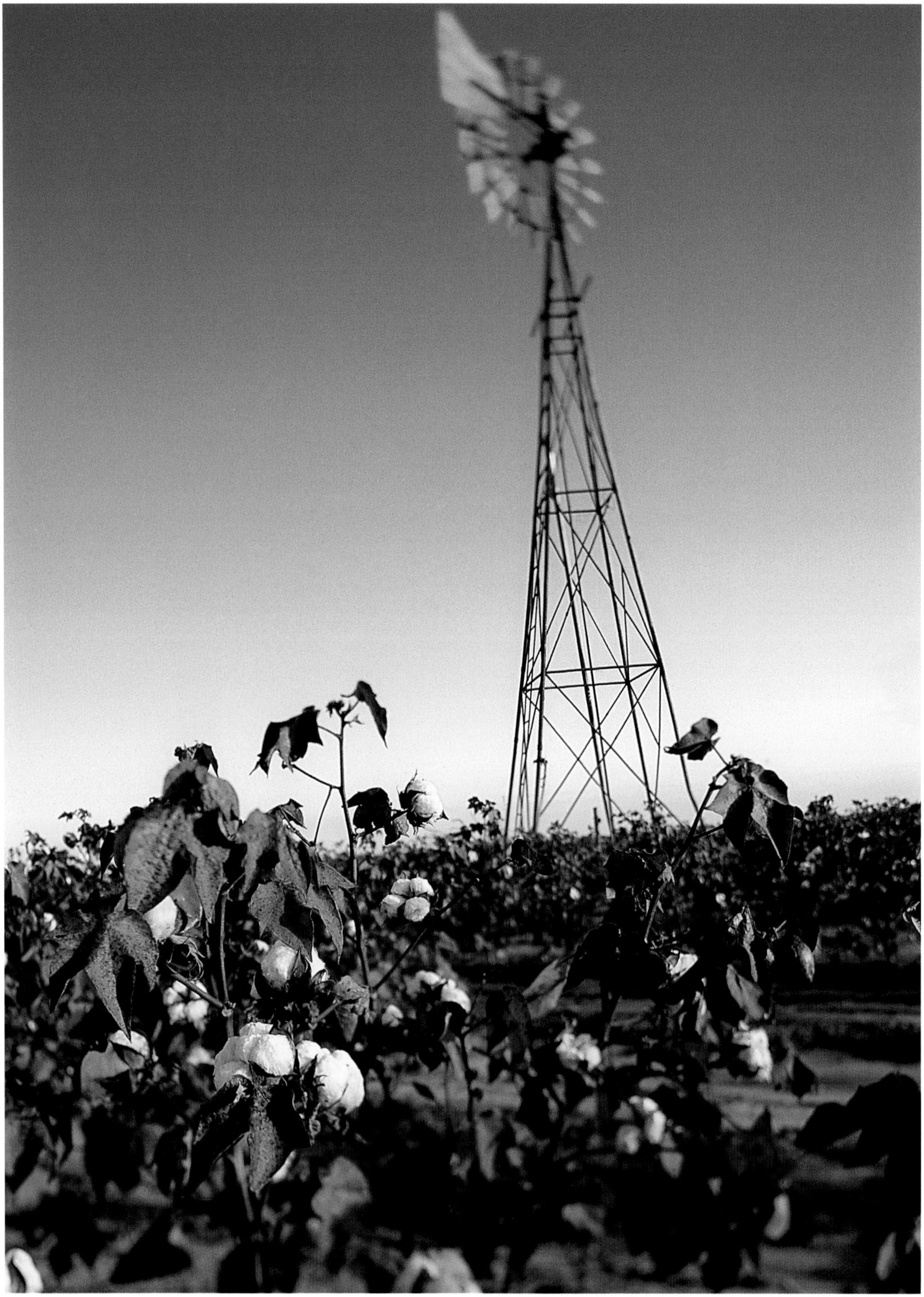

When Bill Miller was a kid, he often visited his granddad's farm. There was a working windmill there, and Bill and his little friends would play near the base of the tower. Bill said they discovered that the well casing, which stuck up above the ground, was a great place to drop things, and plastic army men were the best. "I don't know how many army men we dropped, but it had to have been a lot. I'm sure they are all still down there floating on the water."

A CREW OF WINDMILLERS WERE MAKING REPAIRS to a windmill that took several days of hard work. Usually on these long jobs a cook prepared the food, although he was usually just a windmiller himself. Around noon, two men were up on the tower platform while the others were working on the ground. The cook finished fixing lunch and cried out, "Lunch is on." The men on the ground started toward the chuck wagon, but one of the men on the tower forgot where he was, turned around, and stepped right off the platform. After he bounced on the ground, the cook said, "That's the fastest anyone has ever come to one of my meals."

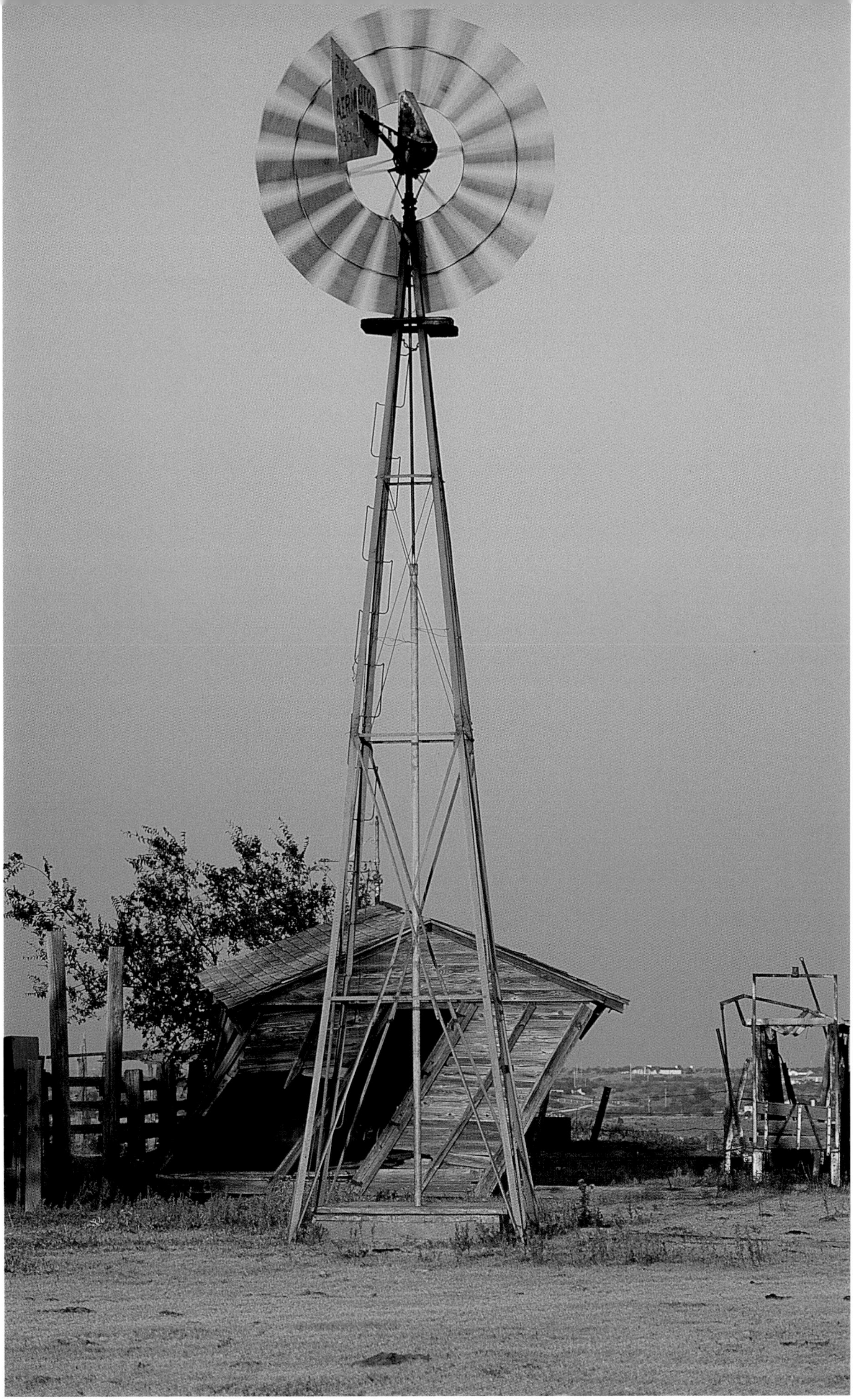
THE
AERMOTOR

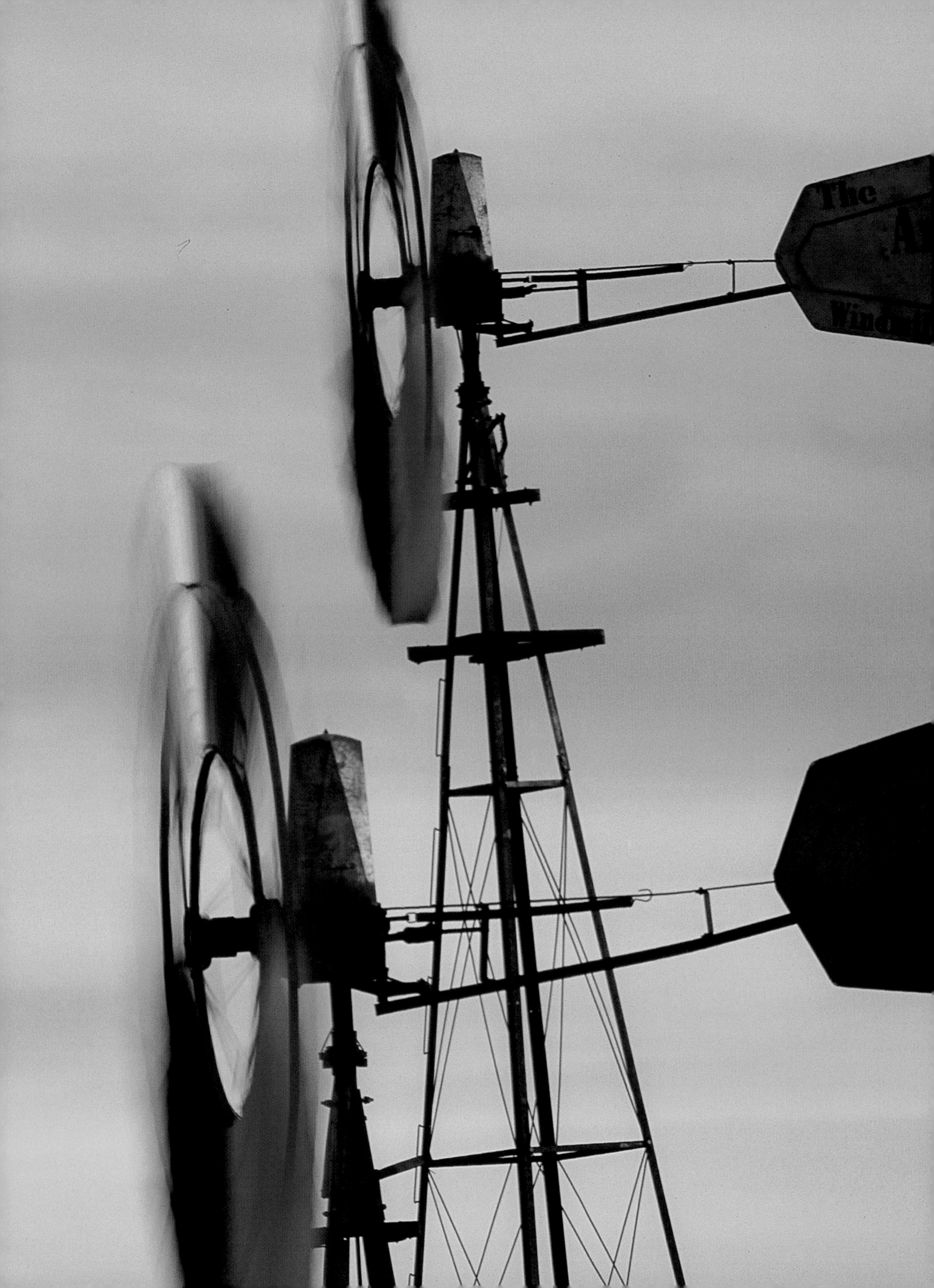
The

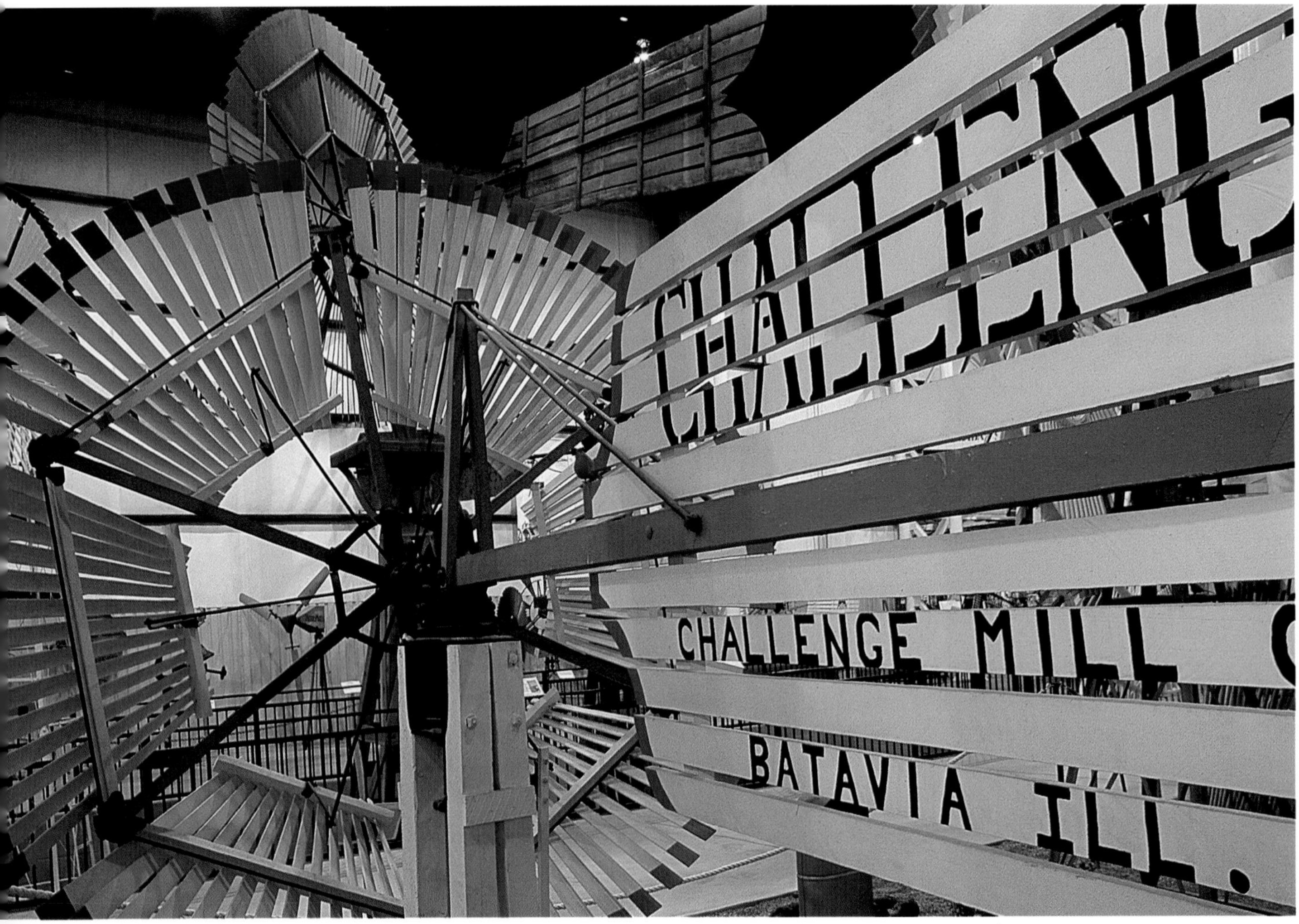
CHALLENGE MILL
BATAVIA ILL.

Acknowledgments

WINDMILL MUSEUM STAFF MEMBERS who collected the windmill stories were Jan Hayes, who recorded the first stories, Rick Nidey, Tanya Meadows, R. G. de Stolfe, Eddie Sosa, Glenn Patton, Dorothy Patton, Carol Anderson, Shelley Harris, Cary Harris, and museum volunteers Suzan Duhan, Calvin Waters, Jeanne Griffin, Scott and Elsie Couch, Glendon Stokes, Alton Brazell, David Simons, Louis Ward, Paul Cowley, Bill Greenlee, Mark Durham, and, of course, Billie Wolfe. Without their attentiveness, these stories would have been lost.

The American Wind Power Center is most grateful to the people who shared their windmill experiences with us. It is surprising how often we hear the same stories from people who lived in different places across America. Most of the stories printed here were told by visitors who did not leave their full names. To those who did—Earl Whitmore, Rick Langston, Carol Davis, Bernice Kahlick, Edward Jennings, Norris Smith, Estelle McCaslin, Siamon Nelson, Robert Lutrick, Vera Surrat, Johnny Williams, Stan Renfro, Carol Reed, Mozelle Wimp, Carolyn Day, Wilbur Ray, Martin Moore, and Bill Miller—thank you very much for the memories.

A special thanks and appreciation is given to Susan Perry, president of the Perry Foundation. Ms. Perry not only provided the funding to produce this book but also lent valuable guidance in its direction and content.

Steve Halladay delightfully agreed to write the introduction to *Windmill Tales.* Steve is related to a pioneering legend in the windmill community, Daniel Halladay, the man who conceived and put into production the first successful American-style windmill.

The trade card illustration came from the library of the American Wind Power Center. Rick Nidey provided the Halladay postcard from his extensive collection.

To all these people and sources sincere thanks is given by the American Wind Power Center.

COY F. HARRIS
Executive Director
August 2003

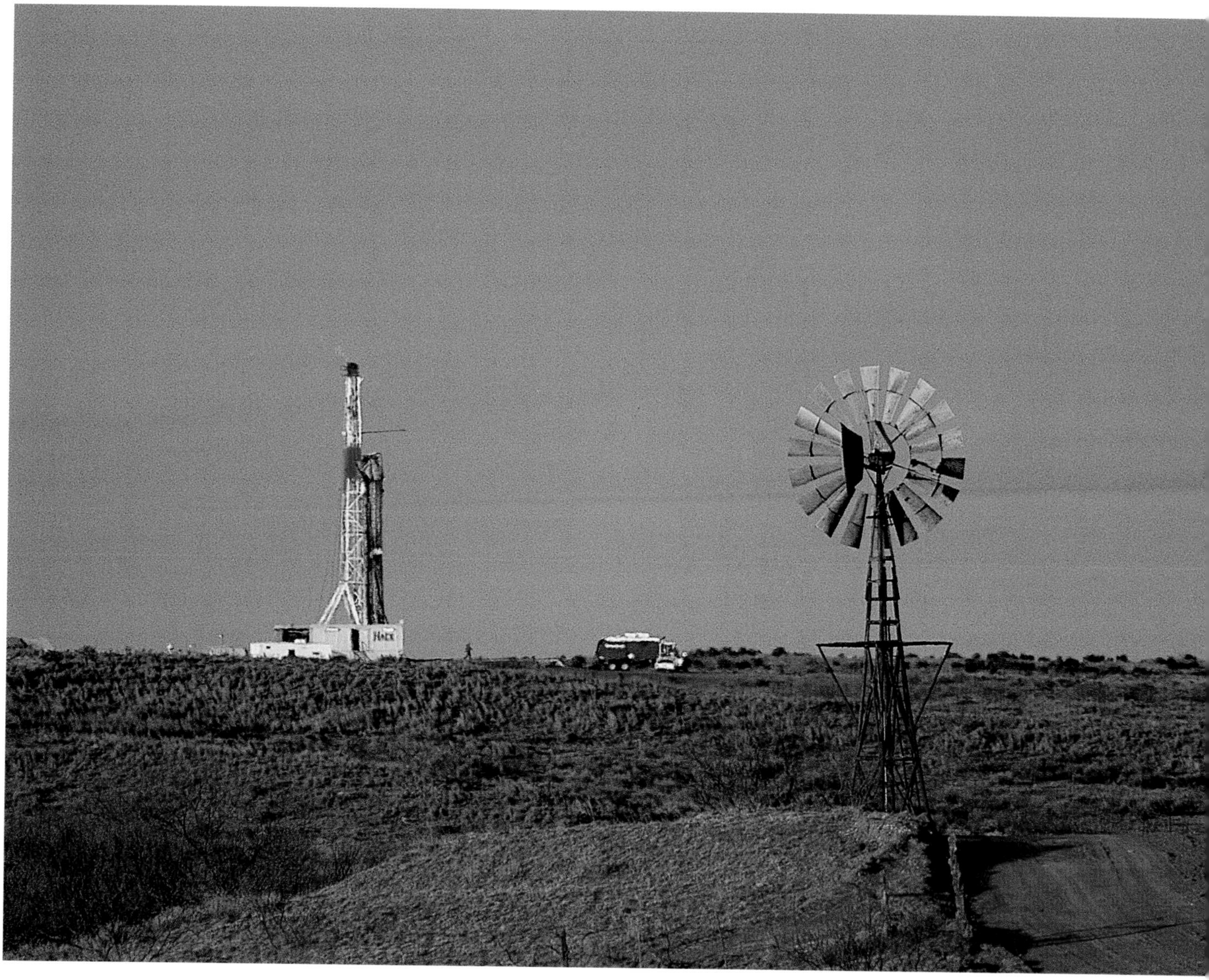

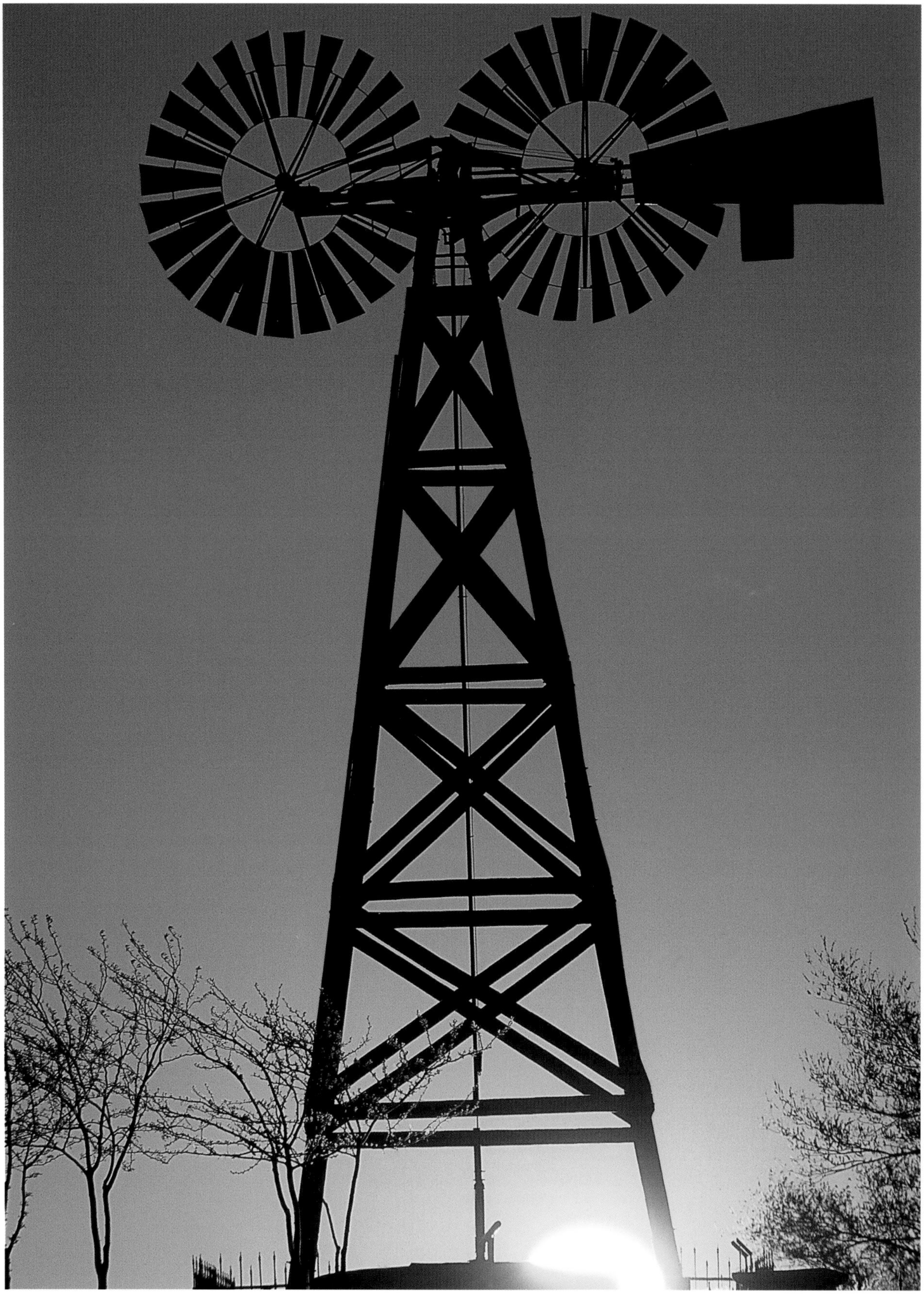

The most unusual Twin Wheel windmill, made in Kansas in the 1920s.

THE AMERICAN WIND POWER CENTER is internationally recognized for its comprehensive collection of historic windmills. Located on twenty-eight acres near downtown Lubbock, it displays over 120 of the machines, many working and pumping water. The rarest of the museum's windmills are exhibited indoors, where visitors can see a variety of early models, from the first all-metal one to the largest and smallest windmills sold in America.

Complementing any visit to the American Wind Power Center are tours of the Windmillers Art Gallery and purchases made in the Windsmith museum store.

The American Wind Power Center
1701 Canyon Lake Drive
Lubbock, Texas 79403
Museum office (806) 747-8734
www.windmill.com

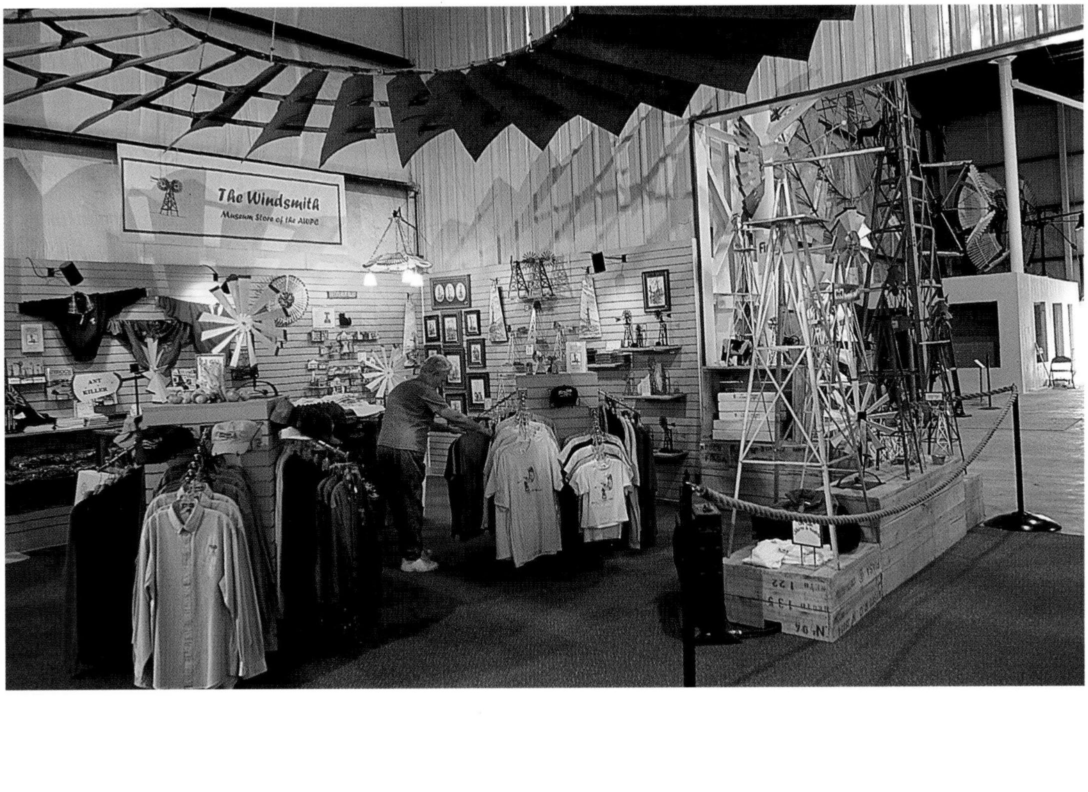
The Windsmith

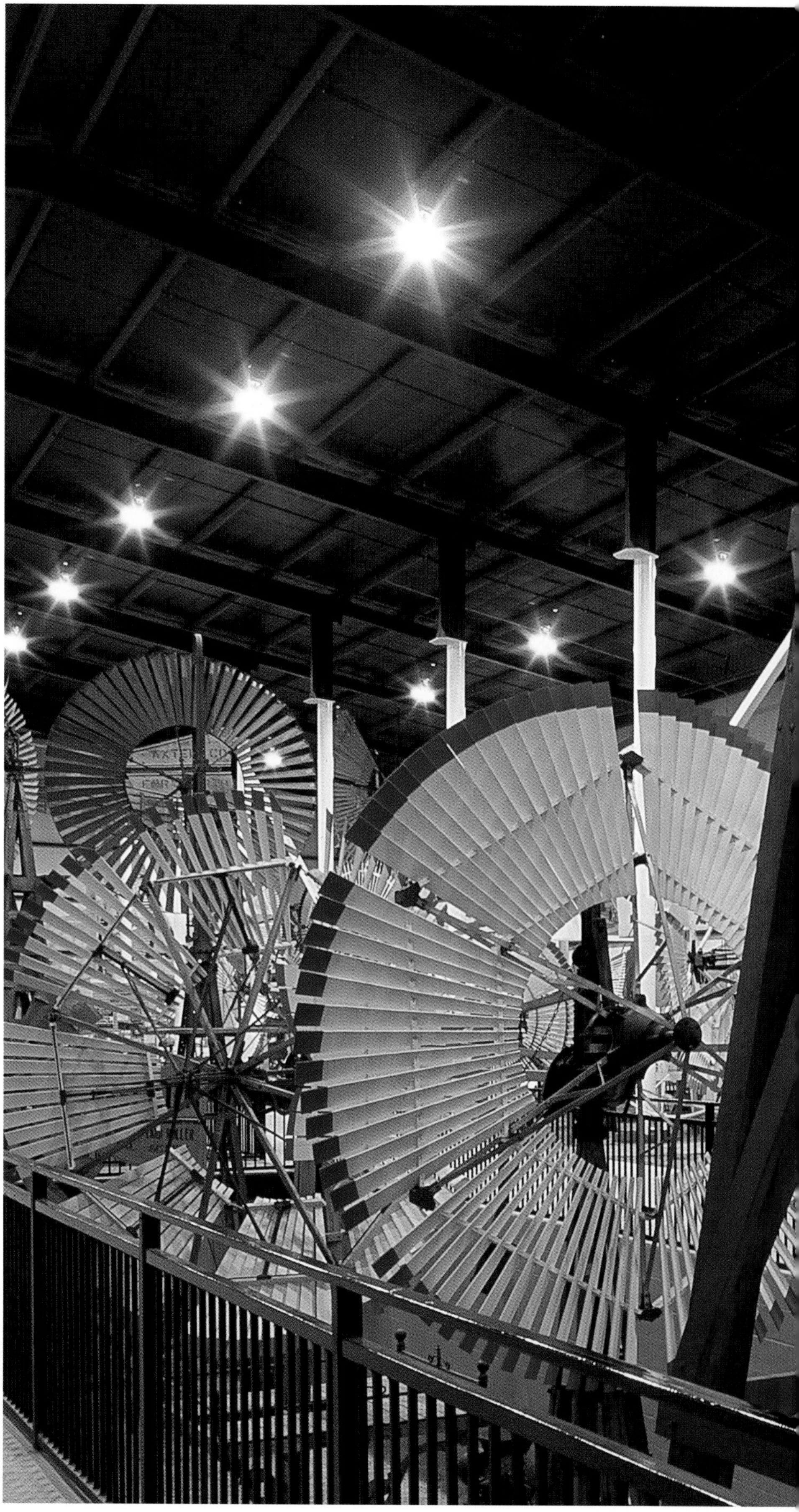

Exhibits in the American Wind Power Center museum.

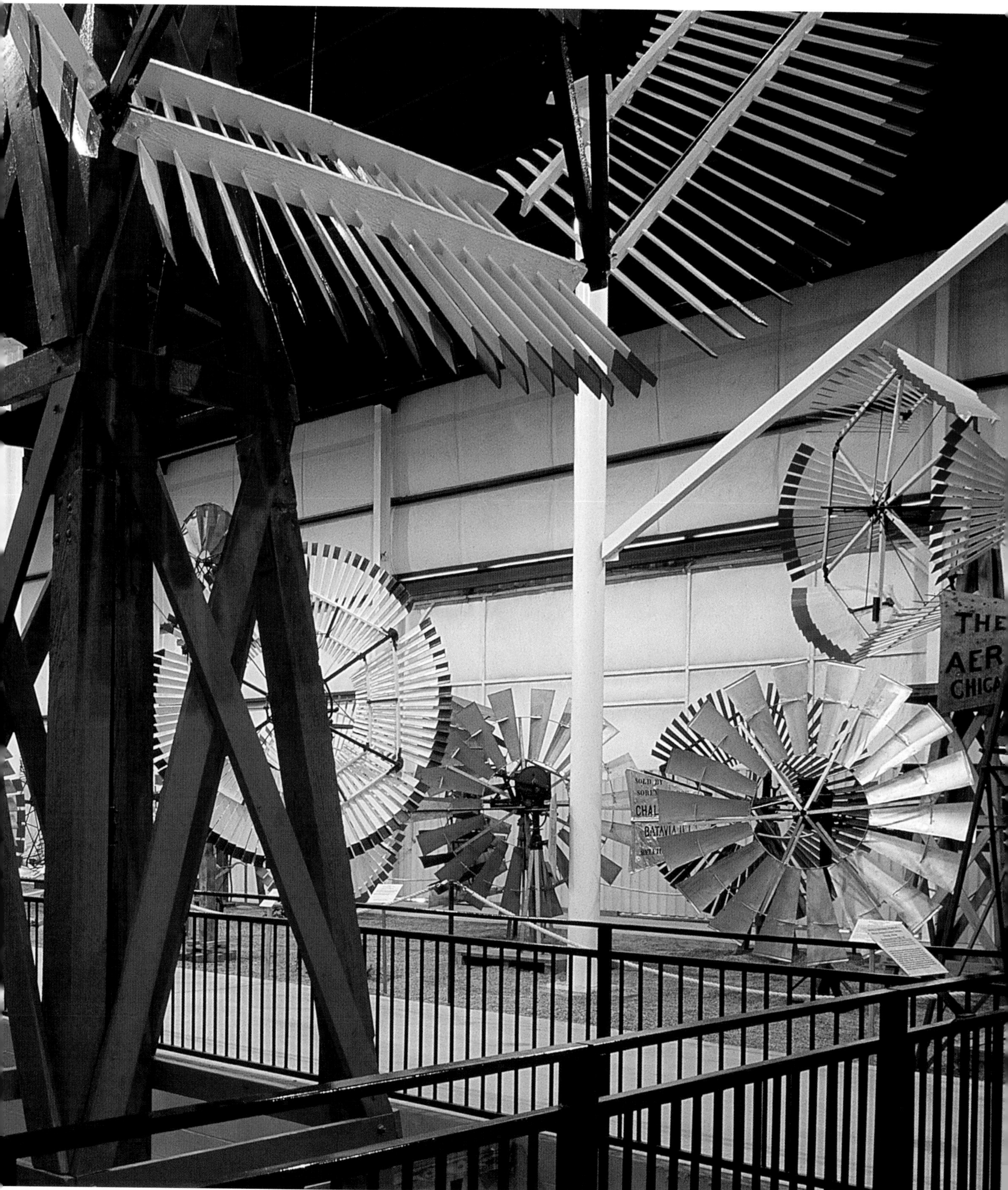
THE
AER
CHICA

The concrete and cast iron battleship weight on a Monitor windmill.

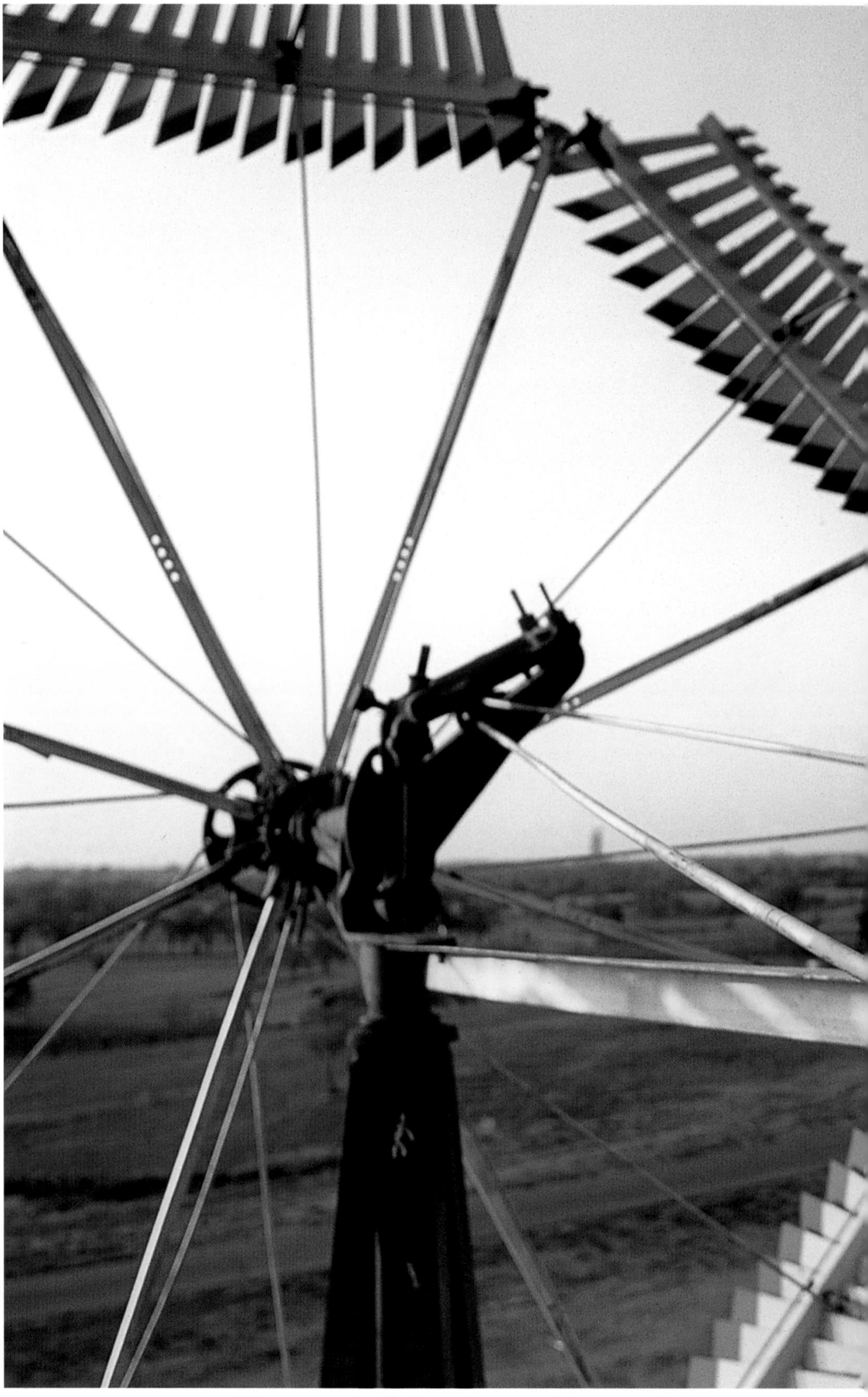

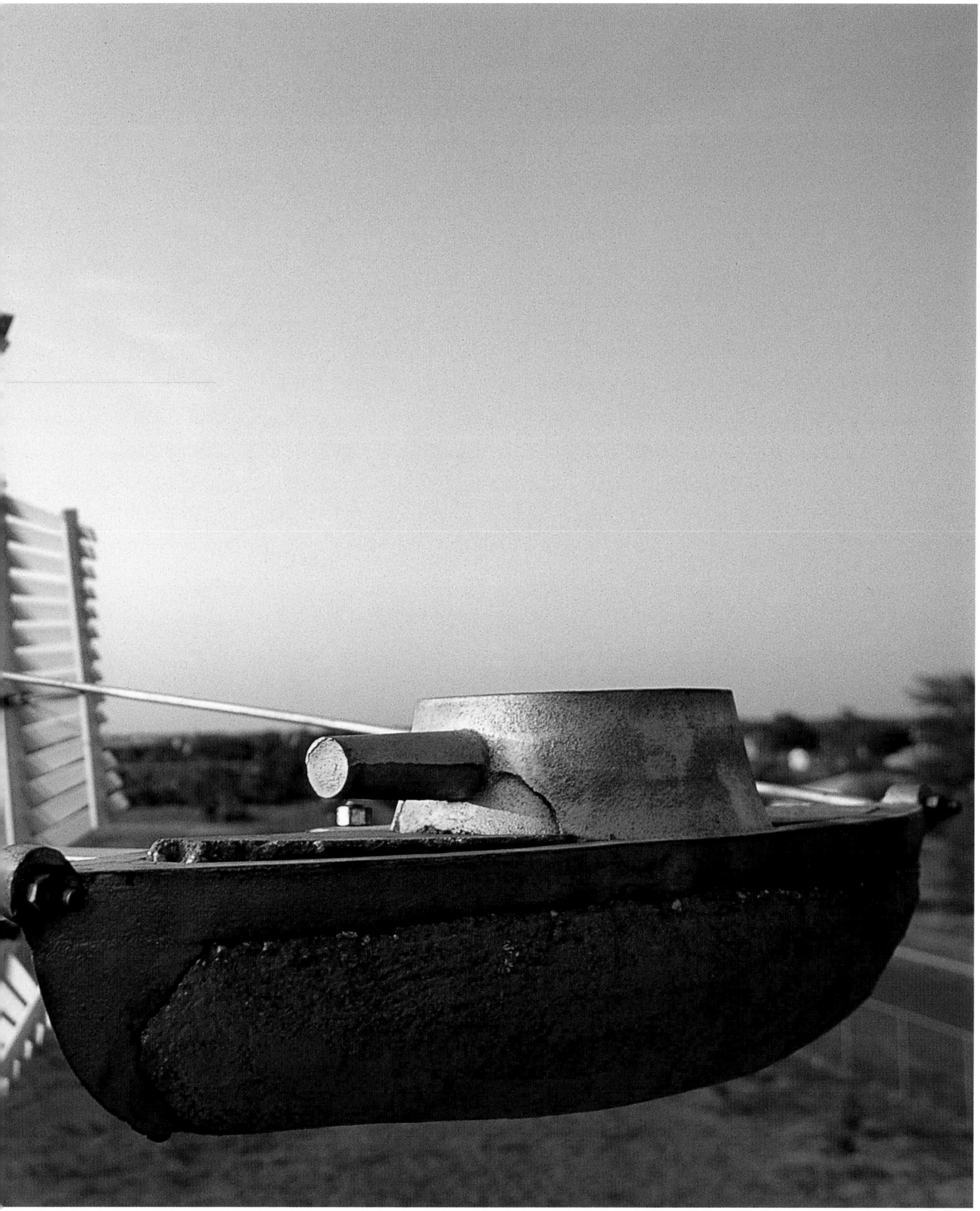

THIS BOOK WAS SET IN MONOTYPE CENTAUR. ORIGINALLY DESIGNED BY BRUCE ROGERS FOR THE METROPOLITAN MUSEUM IN 1914, CENTAUR WAS RELEASED BY MONOTYPE IN 1929. MODELED ON LETTERS CUT BY THE FIFTEENTH-CENTURY PRINTER NICOLAS JENSON, CENTAUR HAS A BEAUTY OF LINE AND PROPORTION THAT HAS BEEN WIDELY ACCLAIMED SINCE ITS RELEASE. THE ITALIC TYPE, ORIGINALLY NAMED ARRIGHI, WAS DESIGNED BY FREDERIC WARDE IN 1925. HE MODELED HIS LETTERS ON THOSE OF LUDOVICO DEGLI ARRIGHI, A RENAISSANCE SCRIBE WHOSE LETTERING WORK IS AMONG THE FINEST OF THE CHANCERY CURSIVES.

Windmill Tales WAS DESIGNED AND TYPESET BY DAVID TIMMONS DESIGN, AUSTIN, TEXAS, AND PRINTED AND BOUND BY SUNG IN PRINTING, KOREA, 2004.

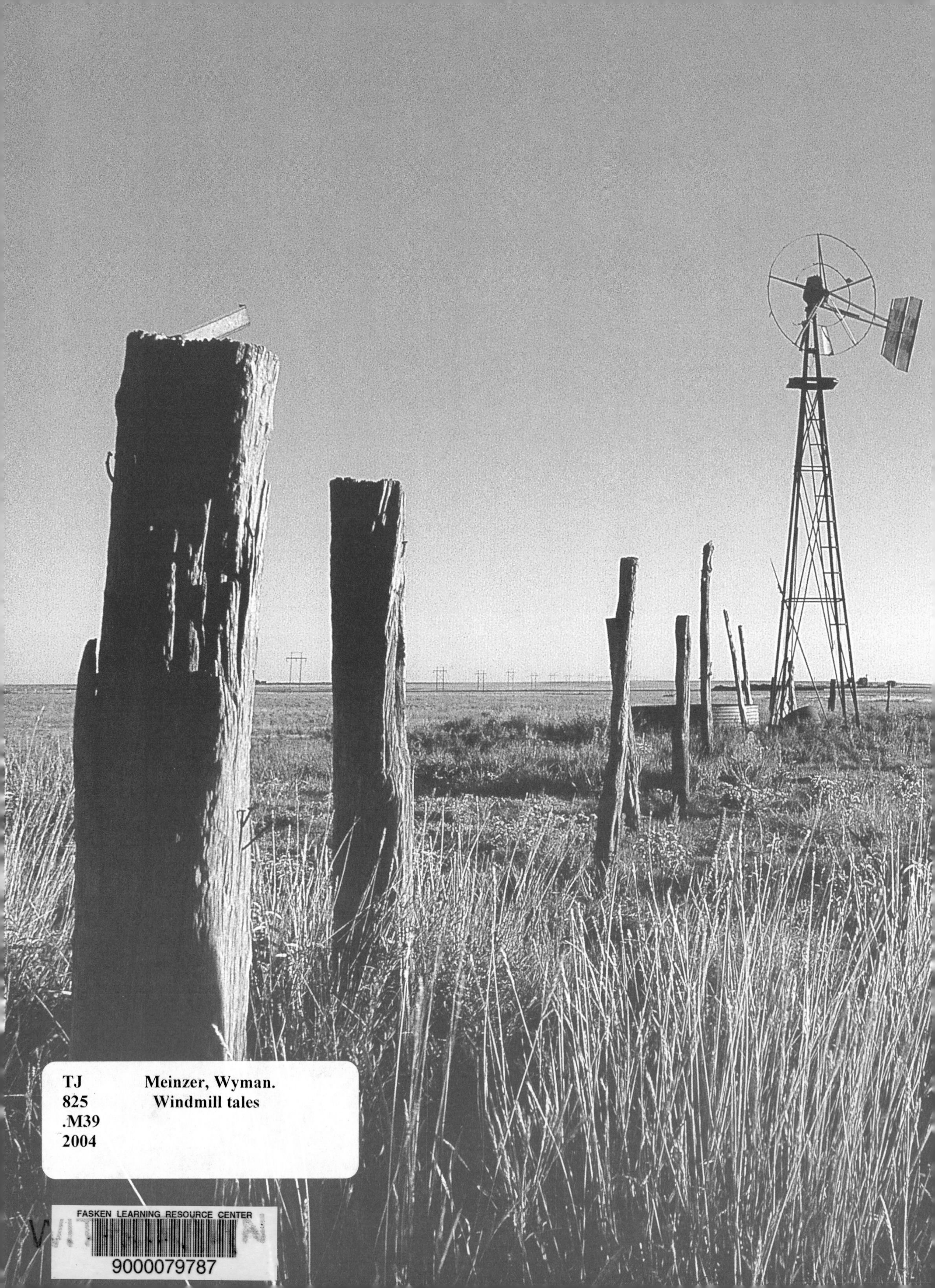